DR. ALAN MELLORS
CHEMISTRY AND BIOCHEMISTRY
UNIVERSITY OF GUELPH
GUELPH, ONTARIO, N1G 2W1 CANADA

Dynamic Models in Biochemistry

A Workbook of Computer Simulations
Using Electronic Spreadsheets

Dynamic Models in Biochemistry

A Workbook of Computer Simulations Using Electronic Spreadsheets

Daniel E. Atkinson

Steven G. Clarke

Douglas C. Rees

The University of California, Los Angeles

David S. Barkley, Editor

Barkley and Associates

The Benjamin/Cummings Publishing Company, Inc.
Menlo Park, California • Reading, Massachusetts •
Don Mills, Ontario • Wokingham, U.K. • Amsterdam •
Sydney • Singapore • Tokyo • Madrid • Bogota •
Santiago • San Juan

Published in association with Barkley and Associates

Sponsoring Editor: Diane Bowen
Production Editor: Merry Finley
Copyeditor: Betty Duncan-Todd
Cover Designer: Dana Chan

Copyright © 1987 by The Benjamin/Cummings Publishing Company, Inc.

All rights reserved. No part of this publication may be reproduced, stored in a retrieval system, or transmitted, in any form or by any means, electronic, mechanical, photocopying, recording, or otherwise, without the prior written permission of the publisher. Printed in the United States of America. Published simultaneously in Canada.

Library of Congress Cataloging-in-Publication Data

Atkinson, Daniel E.
   Dynamic models in biochemistry.

   "Published in association with Barkley and Associates."
   Includes index.
   1. Biological chemistry—Mathematical models.
2. Biological chemistry—Data processing.
3. Electronic spreadsheets. I. Clarke, Steven G.
II. Rees, Douglas C. III. Barkley, David S.
IV. Title.
QP517.M3A85   1987   574.1'92'0285   86-31708
ISBN 0-8053-0420-7

ABCDEFGHIJ-AL-89876

The Benjamin/Cummings Publishing Company, Inc.
2727 Sand Hill Road
Menlo Park, California 94025

We dedicate this book to our teachers and colleagues within the scientific community and within the community of personal computer designers, programmers, and users.

Benjamin/Cummings Titles of Related Interest

R. F. Boyer
Modern Experimental Biochemistry (1986)

L. E. Hood, J. H. Wilson, and W. B. Wood
Molecular Biology of Eucaryotic Cells (1975)

J. D. Watson, N. H. Hopkins, J. W. Roberts, J. A. Steitz, and A. M. Weiner
Molecular Biology of the Gene, Vols. I & II (1987)

W. B. Wood, H. H. Wilson, R. M. Benbow, and L. E. Hood
Biochemistry: A Problems Approach, second edition (1981)

G. Zubay
Biochemistry (1983)

See also

R. Schleif
Genetics and Molecular Biology (1986)

Preface

This book of experiments is itself an experiment. Although personal computers are rapidly finding their way into colleges and universities, the role that personal computers will or should play in education has yet to be defined.

We believe that the personal computer will become a tool for simulation — a device to build and interrogate models of complex events and ideas in disciplines where such simulations have enjoyed only modest successes.

There are at least two possible approaches:

1. Libraries of special purpose programs with extensive graphics displays that provide realistic simulations of physical or biological events.

2. Computer environments within which the student builds the model and where the structure of the model is completely transparent.

This book is an attempt to realize the second of these approaches.

It will be several years before most undergraduate science majors will come to the classroom with even rudimentary programming skills. When this day arrives, probably only a few students will have the talent or desire to spend the hours required to build software models of complex events using procedural languages such as BASIC, Pascal, or C.

However, students can built sophisticated computer simulations by using non-procedural languages. We believe that the best choice of an easy-to-learn, nonprocedural computer language is the electronic spreadsheet (for example, Visicalc, Lotus 1-2-3, Jazz, Excel, Supercalc, Multiplan). Other nonprocedural languages may eventually make their way to the marketplace in a form useful for undergraduate students, but the intuitive man–machine interface that constitutes the electronic spreadsheet will have a place in personal computing for decades to come.

Scope of the Book

This book is designed for students taking a first course in biochemistry. Much of the material drills students on elementary topics. However, personal computers and computer simulations naturally promote independent study and exploration. Rather than cut off an exciting exploration, we have chosen to offer deeply interested students the opportunity to explore some issues in depth. Such pathways can be found in the "Problems" section of each experiment and in experiments marked in the text with an asterisk. Advanced students of biochemistry may find insight and entertainment in these pathways and even in the treatment of elementary topics.

Using the Book

The book is organized as a series of experiments to be performed on personal computers running electronic spreadsheet software. There are no special requirements with respect to the brand of personal computer or electronic spreadsheet program. Electronic spreadsheet graphics capability is convenient but not necessary. Electronic spreadsheets are described in the text using a notation similar to that found in Visicalc, Lotus 1-2-3, Supercalc, Excel, Jazz, and so on (i.e., columns are identified by letters and rows are identified by numbers). Chapters 1 and 2 are designed to help students with no prior computer experience acquire the facility needed to make effective use of this book.

Students build the computer simulation by following the book's detailed directions while sitting at a personal computer. The experiments may be sampled in any order, although a natural progression exists within each chapter.

Each experiment begins with a brief introduction to the quantitative aspects of the topic to be explored and then provides detailed directions for building the electronic spreadsheet model on virtually any personal computer and electronic spreadsheet program. Exercises are divided into a "Questions" section and a "Problems" section.

"Questions" are tightly controlled, and students are held firmly by the hand as they interrogate the model, complete tables, and graph results. "Problems" require varying degrees of student initiative and often permit the interested student to enter domains of genuine, independent exploration.

Experiments that are marked with an asterisk (*) explore advanced topics. However, it is difficult to know what is easy and what is difficult in this new environment. We hope that fundamental concepts that traditionally have been beyond the reach of introductory courses may now be accessible.

This book is a beginning and will grow and prosper only with your help, advice, and comments. Please write us with your experiences as you explore the book.

Acknowledgements

We have benefited from the experience and advice of many people during the four years it has taken to design and write this book.

The idea for the book grew out of discussions with Neil Patterson and Peter Renz at W. H. Freeman & Company and was weaned through its beginnings by Robert Sweet, David Sigman, Verne Schumaker and Richard Weiss at UCLA, Robert Kelley at American McGaw, and Richard Heiser of General Eclectic.

Sandra Lamb directed the early undergraduate testing of the book from her laboratory at UCLA and did much to evangelize the book when such evangelism was most necessary.

Anders Amundson and Integrated Computer Systems helped us to establish an electronic mail link between Los Angeles and London while one of us was on sabbatical.

Geffry Stock and Andrei Lupas at Princeton University gave us the essential final encouragement as the book was prepared for publication.

Jane Gillen and Diane Bowen at Benjamin/Cummings taught us how to get a book published, and Merry Finley showed us how it should look.

The following reviewers provided insight and course correcting at various stages of manuscript development:

    William Bates, The University of North Carolina at Greensboro
    Eleanor Duggan, The University of North Carolina at Greensboro
    Joan Lusk, Brown University
    C. H. W. Hirs, The University of Colorado Health Sciences Center
    William Wood, The University of Colorado, Boulder

David S. Barkley, Editor
Daniel E. Atkinson
Steven G. Clarke
Douglas C. Rees

Contents

Chapter 1		**Getting Started**	1
		The Electronic Spreadsheet	1
		Using Electronic Spreadsheets	4
	1.1	Base Pairing in DNA	6
Chapter 2		**Tips on Using Electronic Spreadsheets**	**10**
	2.1	Entering Labels, Values, and Formulas	10
	2.2	Operator Hierarchy and Parentheses	14
	2.3	Copying Formulas: Relative and Absolute Variables	16
	2.4	Building a Complex Model	19
	2.5	Forward References	26
		Conclusion	29
Chapter 3		**Equilibrium and Acid-Base Relationships**	**31**
	3.1	Simple Equilibrium Relationships	31
	3.2	Acidic Dissociation: The Henderson-Hasselbalch Equation	38
	3.3	Arginine Ionic Forms versus pH	43
	3.4	Buffers	49
	3.5	Partition Between Phases as a Function of pH	55
	3.6	Distribution of Ammonia between Blood and Urine	61
Chapter 4		**Enzyme Kinetics**	**65**
	4.1	Eyring Kinetics and the Boltzmann Distribution	68
	4.2	Assumptions of the Michaelis Treatment	72
	4.3	Simple Michaelis Behavior	80
	4.4	Simple Michaelis Behavior: Catalytic Strength of an Enzyme	88
	4.5	Equilibrium: Competition at a Catalytic Site	91
	4.6	pH Response of an Enzyme	102
	4.7	Cooperative Kinetics: Changes in Substrate Affinity	109
	4.8	Effective K_s	117

	4.9*	An Enzyme with Four Noninteracting Sites	123
	4.10*	Cooperative Kinetics: Variable Degree of Cooperativity	131
	4.11*	Cooperative Kinetics: The Hill Equation	141
	4.12*	Cooperative Kinetics: Modifier Effects	149
	4.13*	Cooperative Kinetics: Stimulation and Inhibition of Reaction Rates by Substrate Analogs	157
Chapter 5		***Metabolism***	**165**
	5.1	Simple Equilibrium: Review and New Considerations	166
	5.2	Multiple Sequential Equilibria	172
	5.3	Kinetic Analysis of Sequential Reactions	177
	5.4	Substrate Concentration as a Function of V_{max}, K_m, and Reaction Velocity	184
	5.5	Competition between Enzymes at a Branch Point	190
	5.6*	Feedback Control of a Biosynthetic Pathway	197
	5.7*	Pairs of Oppositely Directed Metabolic Sequences	209
	5.8*	Redox Reactions in Metabolism	219
Chapter 6		***Membrane Transport***	**227**
	6.1	Passive Transport Through Membranes	229
	6.2	Facilitated Diffusion Through Membranes	241
	6.3	Passive versus Facilitated Transport	248
	6.4*	Active Transport Across Membranes Calculation of $\Delta G'$	253
	6.5*	Active Transport Across Membranes Distribution of Solute	256
	6.6*	Active Transport Across Membranes: Distribution of an Ionic Solute in the Presence of an Electrical Field	260
Chapter 7		***Structure and Stability of Macromolecules***	**266**
	7.1	Three-dimensional Structures of Macromolecules: Restrictions on Torsion Angles	267
	7.2*	Three-dimensional Structures of Macromolecules: Restrictions on Two Consecutive Torsion Angles	277

7	7.3	Electrostatic Influences on Macromolecular Structure: General Considerations	283
	7.4	Electrostatic Influences on Macromolecular Structure: pH Dependence of Protein Stability	294
	7.5	Hydrophobic Effects and Macromolecular Structure	303
		Conclusion	310

Appendices 315

Appendix A: Common Electronic Spreadsheet Formats — 316

Appendix B: Built-in Formulas — 317

Answer Key 319

Index 343

* **Advanced Exercises**

Dynamic Models in Biochemistry

A Workbook of Computer Simulations
Using Electronic Spreadsheets

Chapter 1

Getting Started

Digital computers are *universal* machines that can mimic any discrete machine. Computers can be programmed to simulate the artist's palette, the musician's instrument, the pilot's aircraft, or the biochemist's molecule.

This book will show you how to use your personal computer and an electronic spreadsheet program (such as Lotus 1-2-3 or Multiplan) to simulate the physical and chemical processes that are the foundations of biochemistry. You will have the opportunity to build dynamic models of biochemical events and to explore and experience those events in a simulated laboratory setting. You will be able to observe and manipulate the forces that, for example, control the flux of intermediates along a metabolic pathway or the transport of ions across a membrane.

Computer simulation is a powerful way to gain experience and insight into the nature of difficult ideas or complex relationships. Professional aircraft pilots, for example, log many hours in flight simulators to instill basic skills or to acquire insight into unusual flight situations. Think of this book as a flight simulator for biochemical ideas.

Until recently, computer simulation of physical processes required professional-level programming expertise. The invention of Visicalc, the subsequent development of numerous, sophisticated electronic spreadsheet programs, and the increasing availability of inexpensive personal computers have eliminated this constraint. The digital computer's extraordinary capacity for simulation can be used by almost anyone to sculpt, test, and refine ideas.

The Electronic Spreadsheet

Imagine a large, two-dimensional space divided into an array of rectangular *cells*. Each cell exists at the junction of a single column and row and is uniquely named by the column and row that creates it (Figure 1.1).

Chapter 1

Figure 1.1

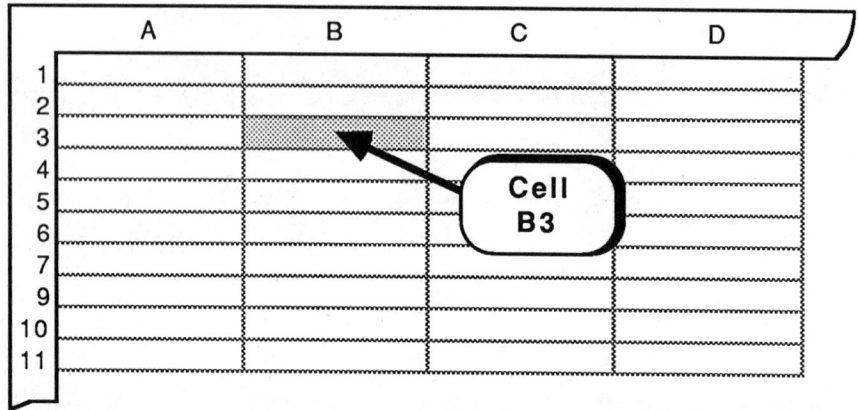

Electronic spreadsheets are organized as large two-dimensional arrays made up of cells. Each cell is uniquely named by the heading of its column and row (e.g., cell B3). In this book we will label columns with letters and rows with numbers. Many, but not all, electronic spreadsheet programs also use this convention. Check yours.

In this book we've adopted the convention of identifying columns with letters (A, B, C, ...) and rows with numbers (1, 2, 3, ...). In Figure 1.1, for example, cell B3 exists at the junction of column B and row 3. Some electronic spreadsheet programs use a different convention (see Appendix A).

Each cell in the array of cells has the potential to be

1. A label
2. A value
3. A formula

A *label* is a string of characters written into a cell to act as a signpost. They enter into no calculations; they influence no results. Often labels identify the contents of a neighboring cell. In Figure 1.2, for example, the label "pH =" in cell A5 points to cell B5 and says "Here lies pH."

A *value* is a number. Cell B1 of Figure 1.2 contains the number 0.05. Values are usually inputs to the computer.

A *formula* defines a calculation. Cell B5 of Figure 1.2 says: "Divide the value of cell B1 by the value of cell B2. Take the logarithm of the result and add to it the value of cell B3." A formula is an algebraic expression

Figure 1.2

	A	B	C
1	[B] =	0.05	
2	[HB] =	0.05	
3	pK =	4.8	
4			
5	pH =	=B3+LOG(B1/B2)	
6			
7			
8			

The cells of electronic spreadsheets may exist as either labels, values, or formulas. Labels are signposts — they tell you where you are. Values are numbers. Formulas are algebraic expressions.

Most electronic spreadsheet programs use a similar syntax (Multiplan is the major exception), but slight differences exist. Built-in functions such as LOG(B1/B2) might be written slightly differently using your electronic spreadsheet program, and formulas may require a symbol other than the preceding "=" in cell B5. Appendix A reviews most of the differences.

whose variables are the names of other cells. Formulas are usually either outputs from the computer or intermediate calculations needed by the formulas of other cells.

Labels, values, and formulas can be arranged together to form *models* of processes. Figure 1.2 describes a model of the pH of an aqueous solution of weak acid. Cell B5 contains an expression for the Henderson-Hasselbalch equation

$$pH = pK + \log([B^-]/[HB]) \qquad (1.1)$$

Figure 1.3a shows how an electronic spreadsheet model might appear on a computer display screen. Cells containing labels and values are displayed without modification. Cells containing formulas are modified to show the computed value of a formula rather than the formula itself. The corresponding web of value- and formula-containing cells gives a continuous readout of the state of the model as the values of individual cells are changed by the user (Figure 1.3b through d).

Figures 1.2 and 1.3 describe a simple model whose consequences might easily be followed with a pocket calculator. The same notation, however, can be used to describe events of almost any complexity. Models described in this notation can serve as laboratories to probe the events that

Chapter 1

Figure 1.3

(a)

	A	B
1	[B] =	0.05
2	[HB] =	0.05
3	pK =	4.8
4		
5	pH =	4.8

(b)

	A	B
1	[B] =	0.05
2	[HB] =	0.02
3	pK =	4.8
4		
5	pH =	5.2

(c)

	A	B
1	[B] =	0.05
2	[HB] =	0.02
3	pK =	3.0
4		
5	pH =	3.4

(d)

	A	B
1	[B] =	0.05
2	[HB] =	0.05
3	pK =	3.0
4		
5	pH =	3.0

The computer display (either the screen or the printer) usually shows the computed value of a formula rather than the formula itself. Thus, as the values of cells are changed, the effect on computed results is seen almost instantaneously.

define living systems. Even relatively simple models can yield instructive lessons.

Using Electronic Spreadsheets

The instruction manual provided with your electronic spreadsheet program quickly shows you how to control your specific program. In general, electronic spreadsheets are controlled in two ways:

1. Point the on-screen cursor to a specific cell of the electronic spreadsheet by using arrow keys on the computer keyboard (or, on some computers, by using a pointing device such as a mouse) and type in the characters that belong to that cell.

2. Issue a command that tells the program to perform a special function such as "print to the line printer" or "save data to the data diskette." Commands are organized as menus of executable options. Some electronic spreadsheets will display a command menu when "/" is typed. Others always display a command menu. Check your manual.

Getting Started

Throughout the rest of this book, you should be sitting next to a personal computer running an electronic spreadsheet program. You should:

1. Read your electronic spreadsheet program instruction manual. Most of the manual is reference material — the essentials are learned in an hour or two.

2. Try a few of the exercises in your instruction manual until you get the "feel" of your electronic spreadsheet program.

3. If you have troubles, talk them over with your local computer wizard. (Somebody you know knows one.)

To get a feel for how an electronic spreadsheet program can be used to model a physical process, try the following exercise.

Chapter 1

Experiment 1.1

Base Pairing in DNA

The two helical, polynucleotide chains that constitute a DNA fiber are held together by hydrogen bonds between pairs of bases. Adenine is always paired with thymine; guanine is always paired with cytosine. Thus, if the adenine content of one of the polynucleotide chains is known, the thymine content of the complementary chain is also known. If the guanine content of one chain is known, the cytosine content of the complementary chain is also known (Figure 1.4).

For example, suppose that strand 1 contains 25% adenine and 35% guanine. Strand 2 must contain 25% thymine and 35% cytosine. It also follows that the remaining 40% of strand 1 must be composed of thymine and cytosine, and the remaining 40% of strand 2 must be composed of adenine and guanine. That is, if the percent adenine content of strand 1 equals x and the percent guanine content of strand 1 equals y, then:

T content (strand 2) = x
C content (strand 2) = y
T + C content (strand 1) = $100 - x - y$
A + G content (strand 2) = $100 - x - y$

Figure 1.4

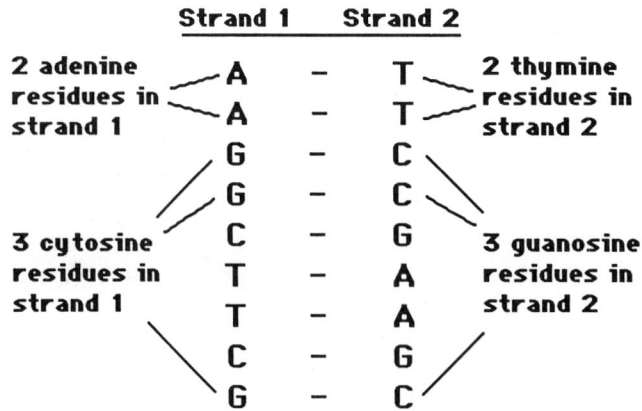

Getting Started

These simple relationships can be used to form the basis of an electronic spreadsheet model that defines the base pair content of a fiber of DNA if the adenine and guanine content of one of the strands is known (Figure 1.5).

Figure 1.5

(a)

Formulas

	A	B	C	D
1	Strand 1		Strand 2	
2	[A] =	30	[T] =	=B2
3	[G] =	24	[C] =	=B3
4	[T + C] =	=100-B2-B3	[A + G] =	=B4

Values

	A	B	C	D
1	Strand 1		Strand 2	
2	[A] =	30	[T] =	30
3	[G] =	24	[C] =	24
4	[T + C] =	46	[A + G] =	46

(b)

Formulas

	A	B	C	D
1	Strand 1		Strand 2	
2	[A] =	30	[T] =	+B2
3	[G] =	24	[C] =	+B3
4	[T + C] =	100-B2-B3	[A + G] =	+B4

Values

	A	B	C	D
1	Strand 1		Strand 2	
2	[A] =	30	[T] =	30
3	[G] =	24	[C] =	24
4	[T + C] =	46	[A + G] =	46

Because different electronic spreadsheet programs have different rules of syntax, this model is displayed twice: (a) and (b).

Many electronic spreadsheet programs have adopted the convention of preceding all formulas with "=" (e.g., cells B4, D2, D3, D4 of [a]). If your program doesn't understand this convention, try replacing "=" with "+" in these cells (b). Most electronic spreadsheets understand one of these two conventions. If your program uses a third convention (e.g. Multiplan), consult your electronic spreadsheet program instruction manual and Chapter 2 of this book.

In this book the conventions shown in (a) will be used to describe models.

Chapter 1

Directions

1. Build the model described in Figure 1.5a (Formulas) by writing the labels, values, and formulas into the appropriate cells. If your electronic spreadsheet program won't accept these entries or gives strange results, try the model described in Figure 1.5b.

2. Verify that the model has been entered correctly by checking the results on your computer display with the figures shown in Figure 1.5a,b (Values).

3. Most electronic spreadsheet programs accept and understand the syntax of one of these alternatives. If your program uses a different convention (e.g., Multiplan), consult your electronic spreadsheet program instruction manual and Chapter 2 of this book.

4. In this book the conventions shown in Figure 1.5a are used to describe models.

5. Use your model to answer the following questions. A pocket calculator and simple algebra could also be used to answer these questions, but use your model to get a taste of what is coming.

Questions

1. If one polynucleotide strand of a DNA fiber contains 24% adenine and 35% guanine, what fraction of the strand consists of either thymine or cytosine?

2. If one polynucleotide strand of a DNA fiber contains 30% adenine and 24% guanine, what fraction of the complementary strand consists or either adenine of guanine?

3. Computer simulations often are used to find answers in the same way that an artillery officer homes in on targets. Trial and error is used to home in on the target. If one polynucleotide strand of a DNA fiber contains 26% adenine, what is the percent guanine of the strand if the percent thymine plus cytosine is 45%? (Yes, an exact arithmetic solution is possible without trial and error changes to cell B3. Such, however, will not always be the case!)

Getting Started

In the next chapter you will conduct a series of "experiments" that shows you how to build models using the descriptions in this book. The experiments give you the opportunity to practice translating the printed material in this book into working computer simulations and give you an opportunity to become comfortable with your electronic spreadsheet program.

Just for fun, part of Chapter 2 builds toward a model of the net electrical charge of a polypeptide at varying pH. If you haven't studied the biochemistry underlying the model, don't worry. Just follow along, plug in the numbers, and watch the model run. The topic is covered in detail in Chapter 3.

Subsequent chapters review the essential features of important biochemical processes and guide you through the construction and interrogation of working models of the material discussed. Additional opportunities for model building will be presented in problem sets within each chapter.

Chapter 2

Tips on Using Electronic Spreadsheets

Electronic spreadsheet programs are designed and written by companies that want to sell their products to the largest possible audience. They are designed to be easy to use. However, even if you are familiar with electronic spread-sheet programs and their operation, you may want to skim this chapter quickly to become familiar with the format of this book and the way it presents electronic spreadsheet models.

Experiment 2.1

Entering Labels, Values, and Formulas

All electronic spreadsheets work essentially the same way. They

1. Point to a cell
2. Write into a cell
3. Repeat on a new cell

Unfortunately, the rules governing syntax for the contents of a cell are not consistent from program to program. In this experiment you will identify some of the specific syntax requirements of your program and compare them with the syntax that is used throughout this book.

1. Does your electronic spreadsheet program require you to preface all labels with a double quote (") (e.g., Supercalc), or does it automatically assume that entries beginning with a letter of the alphabet are labels (e.g., Jazz, Lotus 1-2-3)?

2. Are formulas identified by a preceding "=", or are formulas identified by some other mechanism. (Lotus 1-2-3 formulas, for example, begin with a number or with an operator such as

Entering Labels, Values, and Formulas

"+" or "−". Supercalc assumes that any entry that begins with a letter of the alphabet is a formula.)

In this book we will precede formulas with "=", as in:

=B3+B5

3. Do built-in functions such as LOG, SIN, EXP require a preceding "@" (e.g., Lotus 1-2-3), or is this symbol forbidden (e.g., Supercalc, Jazz)?

4. Is LOG(B1/B2) understood by your program (e.g., Jazz) or is the logarithm of base 10 expressed differently in your program (e.g., LOG10(B1/B2) for Supercalc)?

Logarithms seem to be the biggest offender in most programs. Other built-in functions tend to retain their form across different electronic spreadsheet programs (except, of course, for the requirement or refusal of "@"). This book uses LOG for log of base 10 and LN for log of base *e*.

The above list identifies most of the major differences among electronic spreadsheet programs. If your program has others you will probably find them now, as you build the following model.

You might find that your first few sessions at the keyboard are frustrating. There are only a few things to learn, but they all seem to pile up at the beginning.

Directions

1. Review the instruction manual of your electronic spreadsheet program.

2. Build the model in Figure 2.1 on your computer by pointing the on-screen cursor to the appropriate cell and typing in a label, value, or formula that conforms to the syntax requirements of your program. Appendix A of this book may help.

Chapter 2   Experiment 2.1

Figure 2.1

Formulas

	A	B
1	B =	0.05
2	HB =	0.02
3	pK =	4.8
4		
5	pH =	=B3+LOG(B1/B2)

Values

	A	B
1	B =	0.05
2	HB =	0.02
3	pK =	4.8
4		
5	pH =	5.1979

Figure 2.1 is divided into two parts: Formulas and Values. This pattern is followed throughout the book. Use Formulas to build your model; use Values to verify its correctness.

3. If you run into trouble, watch the screen as you type in the offending characters — helpful error messages may appear. Also, ask yourself what the model is trying to do (e.g., take a logarithm of base 10) and then look in the index of your electronic spreadsheet instruction manual for your program's rules in performing this task.

4. Save the model to a data diskette, if you wish.

5. Answer the questions at the end of the exercise.

Questions

1. Examine the model on your computer screen. Does it correspond to Figure 2.1 (Values)? If it does, you have probably entered the model successfully. If it doesn't, you either have not achieved a successful entry, or you are seeing differences in display format (i.e., 5.1979 vs. 5.2). Ignore display format differences now and throughout this book.

2. Does your program assume that any character string that begins with a letter is a label? If not, how does your program identify labels? (Some of the newest electronic spreadsheet programs have very sophisticated tests.)

Entering Labels, Values, and Formulas

3. How does your program identify formulas? A preceding "="? Letters of the alphabet? Numbers or operators (e.g., 0 ... 9, +, –)?

4. Does your program assume that all character sequences that begin with numbers are values? What about special characters such as ".", "(", "@"?

5. What built-in functions does your program support? Compare it with the list of functions in Appendix B. Are there any syntax differences? Write them down for later reference.

6. Cell A1 contains a label that identifies the concentration of the base A. A more informative label for cell A1 might be "(B) =," where "(B)" means "the concentration of B." Attempt to change cell A1 to this new form. If you cannot it is probably because your program thinks you're trying to enter a value (see Question 4, above). In most electronic spreadsheet programs a double or single quote (" or ') forces a label. Try it.

Experiment 2.2

Operator Hierarchy and Parentheses

Some electronic spreadsheet programs work their way through a formula, from left to right, without considering operator precedence. For example, consider the formula

=A1+ A2*A3

Most versions of Visicalc would work through this formula from left to right by first taking A1, adding A2 to A1, and finally multiplying A3 by the previous result (+A1+A2).

Subsequent electronic spreadsheet programmers added "intelligence" to their programs by taking into account the operator precedence rules of conventional algebra (e.g., multiply factors before adding terms). Supercalc or Lotus 1-2-3 would take =A1+A2*A3 and first multiply A2 by A3 and then add A1 to the result.

For most values of A1, A2, and A3 these two procedures would give different answers.

In this book all formulas are written so that they give the same answer no matter what program you are using. For example:

=A1+(A2*A3)

Nevertheless, to write your own models you need to know the rules for your specific program. This experiment will help you discover these rules.

Directions

1. Enter the model described in Figure 2.2 into your computer. The Henderson-Hasselbalch equation has been rearranged to compute the ratio of base to acid at a given pK and pH. The fraction of ionizable group as base is also computed.

Operator Hierarchy and Parentheses

Figure 2.2

Formulas

	A	B
1	pK =	4.8
2	pH =	4.7
3		
4	[B]/[HB] =	=(10^B2)/(10^B1)
5	Frtn =	=B4/(1+B4)

Values

	A	B
1	pK =	4.8
2	pH =	4.7
3		
4	[B]/[HB] =	0.79
5	Frtn =	0.44

Note: "^" means "raised to the power of" in most systems.

The equations are:

$$\text{Fraction} = \frac{[B]/[HB]}{1 + [B]/[HB]} \tag{2.1}$$

$$[B]/[HB] = 10^{pH}/10^{pK} \tag{2.2}$$

2. Save the model to a data diskette if you wish.

3. Answer the questions at the end of the exercise.

Questions

1. Examine the model on your computer screen. Does it correspond to Figure 2.2 (Values) except for minor differences in format?

2. Change the contents of cell B4 to read:

   =10^B2/10^B1

   Does the computed value of this formula remain the same (i.e., 0.79) or does it change? If it changes, what order of calculation leads to the new result?

3. On the basis of your results in question 2, does your electronic spreadsheet program support the rules of operator precedence? If yes, check your program instruction manual for the rules used.

Experiment 2.3

Copying Formulas: Relative and Absolute Variables

Although the cycle of pointing to a cell, writing into a cell, and repeating on a new cell ultimately builds a model of any complexity, your electronic spreadsheet program has features that ease the task of building large models. You will want to learn about all of them by reading your program's instruction manual. In this experiment you will learn about one feature that you will use frequently: The "Copy" (or "Replicate") command.

Directions

1. Examine the model described in Figure 2.3. The model displays the ratio of conjugate base to conjugate acid and the fraction of the species that exists as conjugate base for a range of pH values from 2 to 10.

   Notice that the model is really only repeated copies of the model described in Figure 2.2 — arranged as a table and replicated down the page.

Figure 2.3

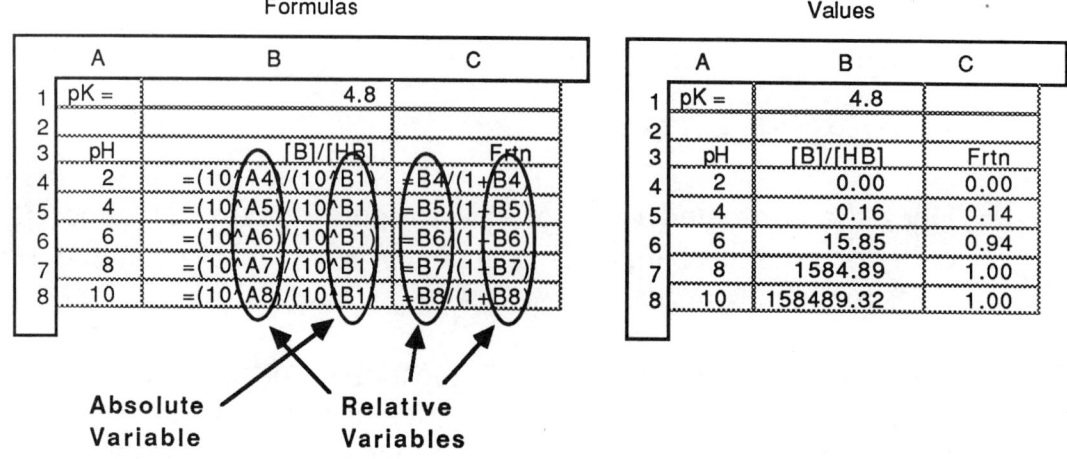

Copying Formulas: Relative and Absolute Variables

The formulas in column B, for example, are copies of the Henderson-Hasselbalch equation, $[B]/[HB] = (10^{pH}/10^{pK})$, where pH varies and pK is held constant. The variable that points to values of pH is called a *relative variable*. As the formula changes location from cell B4 to B8, the variable changes from A4 to A8. The variable that points to the value of pK is called an *absolute variable*. This variable always points to cell B1 no matter where the formula that contains the variable is located.

Relative variables point to cells relative to the location of the formula (e.g., "one to the left"). Absolute variables point to specific cell locations such as B1.

Your electronic spreadsheet instruction manual describes the behavior of relative and absolute variables in detail.

2. Electronic spreadsheet programs are capable of copying formulas with both relative and absolute variables. Thus you only need type the formula once. The copying routine of your program takes care of the rest. Go to your electronic spreadsheet instruction manual now and study the Copy (sometimes called Replicate) command of your program.

    Build the model described in Figure 2.3 by using the Copy command of your program.

3. In this book we sometimes use a special shorthand to describe multiple copies of formulas that contain relative and absolute variables. We do this when there is no room to conveniently write the formula in the matrix of the model.

    Examine Figure 2.4. Absolute variables are highlighted in **bold** print. Notice that you only really need to see the formula once. If you know the formula in cell B4, for example, and you know which variables in the formula are relative and which are absolute, you can write the formulas in cells B5 through B8. If the formulas were long and there were no room to place them in the matrix of the model itself, the model could be described as in Figure 2.5.

    Study Figures 2.4 and 2.5 until you understand how one may be built from the other. Throughout this book you will find models described as in Figure 2.5. This saves room and helps you make

Chapter 2   Experiment 2.3

better use of the Copy command of your electronic spreadsheet program.

Figure 2.4

Formulas

	A	B	C
1	pK =	4.8	
2			
3	pH	[B]/[HB]	Frtn
4	2	=(10^A4)/(10^B1)	=B4/(1+B4)
5	4	=(10^A5)/(10^B1)	=B5/(1+B5)
6	6	=(10^A6)/(10^B1)	=B6/(1+B6)
7	8	=(10^A7)/(10^B1)	=B7/(1+B7)
8	10	=(10^A8)/(10^B1)	=B8/(1+B8)

Values

	A	B	C
1	pK =	4.8	
2			
3	pH	[B]/[HB]	Frtn
4	2	0.00	0.00
5	4	0.16	0.14
6	6	15.85	0.94
7	8	1584.89	1.00
8	10	158489.32	1.00

Figure 2.5

Formulas

	A	B	C
1	pK =	4.8	
2			
3	pH	[B]/[HB]	Frtn
4	2	Formula Set I	Formula Set II
5	4		
6	6		
7	8		
8	10		

Values

	A	B	C
1	pK =	4.8	
2			
3	pH	[B]/[HB]	Frtn
4	2	0.00	0.00
5	4	0.16	0.14
6	6	15.85	0.94
7	8	1584.89	1.00
8	10	158489.32	1.00

Formula Set I
Prototype Cell is B4
=(10^A4)/(10^B1)

Formula Set II
Prototype Cell is C4
=B4/(1+B4)

Experiment 2.4

Building a Complex Model

Let's consolidate your knowledge of electronic spreadsheets by building a model of the ionization properties of a small protein.

You may not have studied this material in your biochemistry course yet, but that's all right. In this experiment you're really studying electronic spreadsheets.

Proteins are made up of one or more chains of covalently linked amino acids. In part, the acidic properties of a protein are simply the sum of the acidic properties of each amino acid side group (plus the terminal $-NH_3^+$ and $-COO^-$ groups). Perhaps you already know that neighboring amino acids can interact with each other to complicate the situation.

Barring the interactions that occur between neighboring amino acids, you can build an approximation to the acidic behavior of a protein by writing a Henderson-Hasselbalch equation for each ionizable group in the side chain of each amino acid in the protein and then summing the results.

Directions

1. Build the model described in Figure 2.6 by using the Copy command of your electronic spreadsheet and by keeping track of relative and absolute variables.

   You may find that keyboard entry slows down considerably as the model increases in size. This is because every change you make in the growing model causes the entire model to recalculate. You can turn off this automatic recalculation feature and have the model recalculate only under your command. Check your program instruction manual to find out how to do this. *When automatic recalculation is off, request a recalculation before reading results.*

   Verify the correctness of your model by checking your results against the values displayed in Figure 2.6 (Values). As usual, you

Chapter 2   Experiment 2.4

can ignore differences that exist because of differences in display format (e.g., number of decimals showing).

2. By trial and error find a pH value that leads to a net positive charge, a net negative charge, and a zero net charge. (pH = 8.9 yields a zero net charge. pH > 8.9 yields net negative charge. pH < 8.9 yields net positive charge.

Figure 2.6

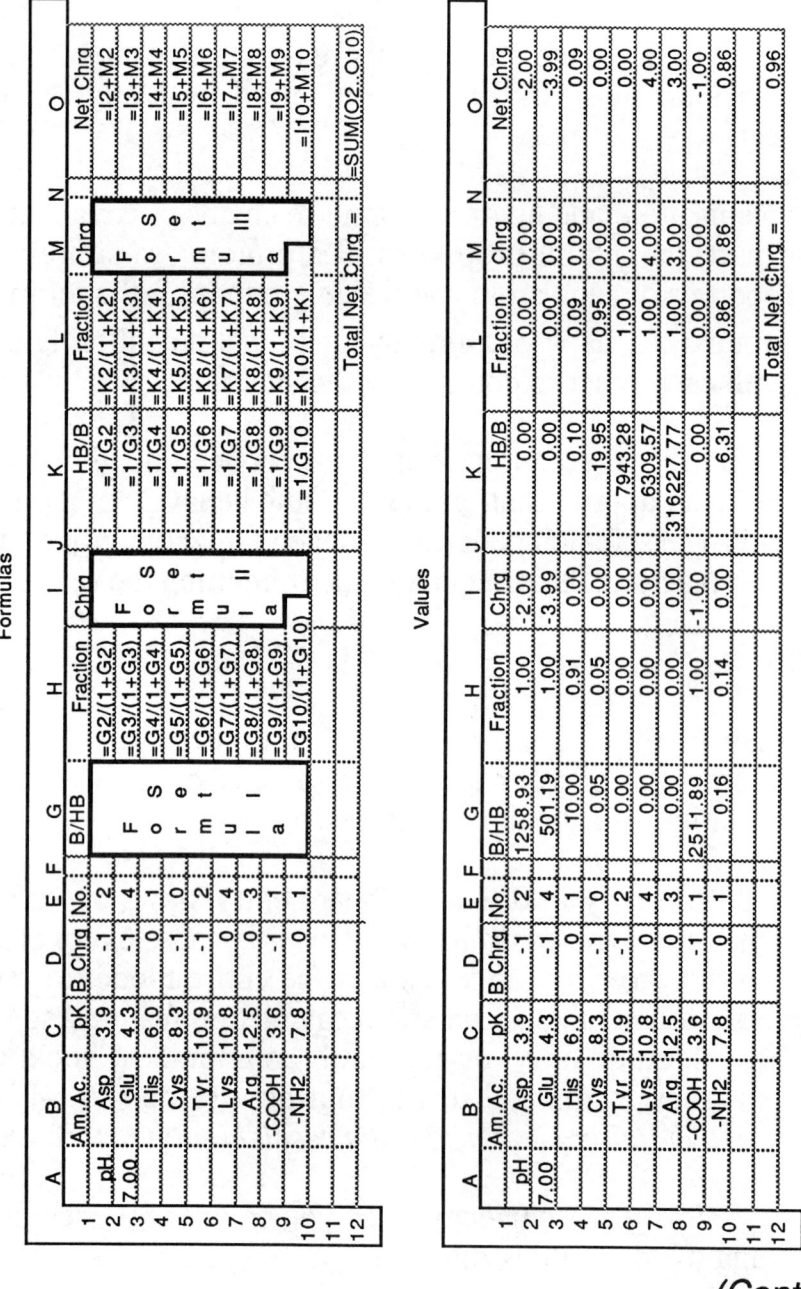

(Continued.)

Figure 2.6 (Contd.)

Formula Set I
Prototype Cell is G2
=(10^**A3**)/(10^C2)

Formula Set II
Prototype Cell is I2
=D2*E2*H2

Formula Set III
Prototype Cell is M2
=(1+D2)*E2*L2

3. A few of the more advanced models in this book (advanced models are marked with an asterisk *) use a more sophisticated notation for relative and absolute variables.

   If you plan to do these exercises you should understand about variables that behave *relatively* in one direction and *absolutely* in the other. Suppose that you wanted to study the effect of both pH and pK on the ratio of conjugate base to conjugate acid. Then you might build a model such as that described by Figure 2.7. The variable that "looks up" pH (A2 in the formula in cell B2) behaves relatively when it is copied down a column, but behaves absolutely when it is copied across a row (i.e., in the variable A2, A is absolute but 2 is relative). Similarly, the variable that "looks up" pK (B1 in the formula in cell B2) behaves absolutely when it is copied down a column, but behaves relatively when it is copied across a row (i.e. B is relative but 1 is absolute). Such behavior may be represented in our previous notation if the column letters and row numbers are tracked separately. The model described in Figure 2.7 may then be described as in Figure 2.8.

Chapter 2 Experiment 2.4

Figure 2.7

Formulas

	A	B	C	D	E
1	pH pK	2	4	6	8
2	2	=(10^A2)/(10^B1)	=(10^A2)/(10^C1)	=(10^A2)/(10^D1)	=(10^A2)/(10^E1)
3	4	=(10^A3)/(10^B1)	=(10^A3)/(10^C1)	=(10^A3)/(10^D1)	=(10^A3)/(10^E1)
4	6	=(10^A4)/(10^B1)	=(10^A4)/(10^C1)	=(10^A4)/(10^D1)	=(10^A4)/(10^E1)
5	8	=(10^A5)/(10^B1)	=(10^A5)/(10^C1)	=(10^A5)/(10^D1)	=(10^A5)/(10^E1)
6	10	=(10^A6)/(10^B1)	=(10^A6)/(10^C1)	=(10^A6)/(10^D1)	=(10^A6)/(10^E1)

Figure 2.8

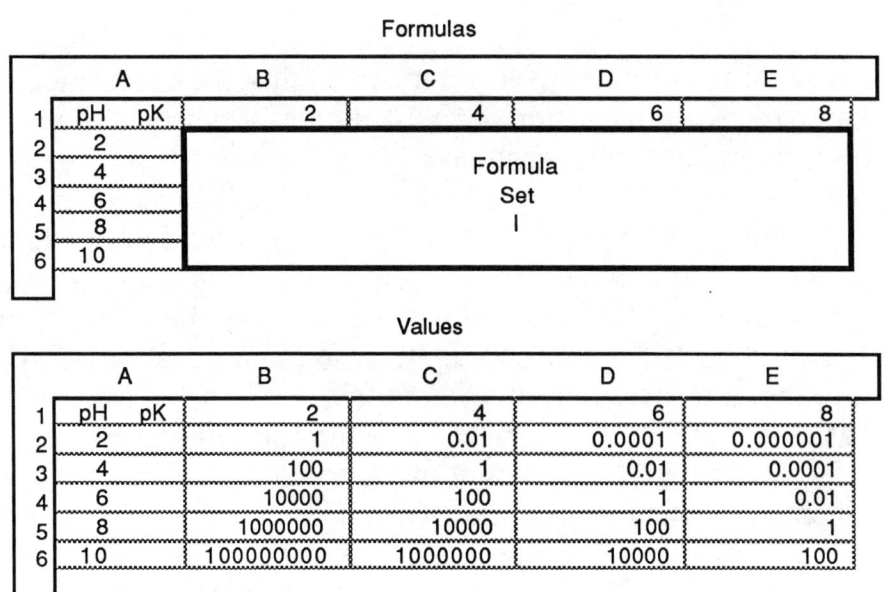

Formulas

	A	B	C	D	E
1	pH pK	2	4	6	8
2	2				
3	4		Formula		
4	6		Set		
5	8		I		
6	10				

Values

	A	B	C	D	E
1	pH pK	2	4	6	8
2	2	1	0.01	0.0001	0.000001
3	4	100	1	0.01	0.0001
4	6	10000	100	1	0.01
5	8	1000000	10000	100	1
6	10	100000000	1000000	10000	100

Formula Set I
Prototype Cell is B2
=(10^**A**2)/(10^**B**1)

4. Build the model described in Figure 2.8 (Formulas), and verify that it behaves correctly by comparing it with Figure 2.8 (Values).

You will have to become very familiar with the Copy command of your electronic spreadsheet program (and with the concepts of relative and absolute variables) to succeed. If you have trouble, put this problem aside for awhile and come back to it another time.

Building a Complex Model

Questions

1. An important feature of electronic spreadsheets is the ability to track the consequences of strings of simple calculations. The Henderson-Hasselbalch equation, for example, is a simple calculation with (after a little thought) intuitively obvious consequences. The consequences of the calculations described in Figure 2.6 are not so intuitive even though they constitute little more than a sequence of Henderson-Hasselbalch calculations and some arithmetic to tie it together. Nevertheless, by playing with the model described in Figure 2.6 you can gradually come to have a "feel" for the modeled polypeptide's behavior.

    Intuition is often sparked by the ability to ask of an object: "What if I do this to you, how will you respond?" This is true of molecule objects, mathematical objects, people objects, and utterly unknown objects.

    Although you may not have a complete understanding of the biochemistry that supports the model described in Figure 2.6, perhaps you're anxious to start asking interesting "what if" questions of interesting models. For example, use Figure 2.6 to verify the table below and Figure 2.9.

Total Net Charge

	No. Asp	2	5	8	2	2	2
pH	No. Tyr	2	2	2	2	5	8
2		8.93	8.89	8.86	8.93	8.93	8.93
4		5.82	4.15	2.48	5.82	5.82	5.82
6		1.58	-1.39	-4.37	1.58	1.58	1.58
8		0.39	-2.61	-5.61	0.39	0.39	0.38
10		-0.77	-3.77	-6.77	-0.77	-1.11	-1.44
12		-6.34	-9.34	-12.34	-6.34	-9.12	-11.89

* The numbers of all other residues are as displayed in Figure 2.6.

Chapter 2   Experiment 2.4

Figure 2.9

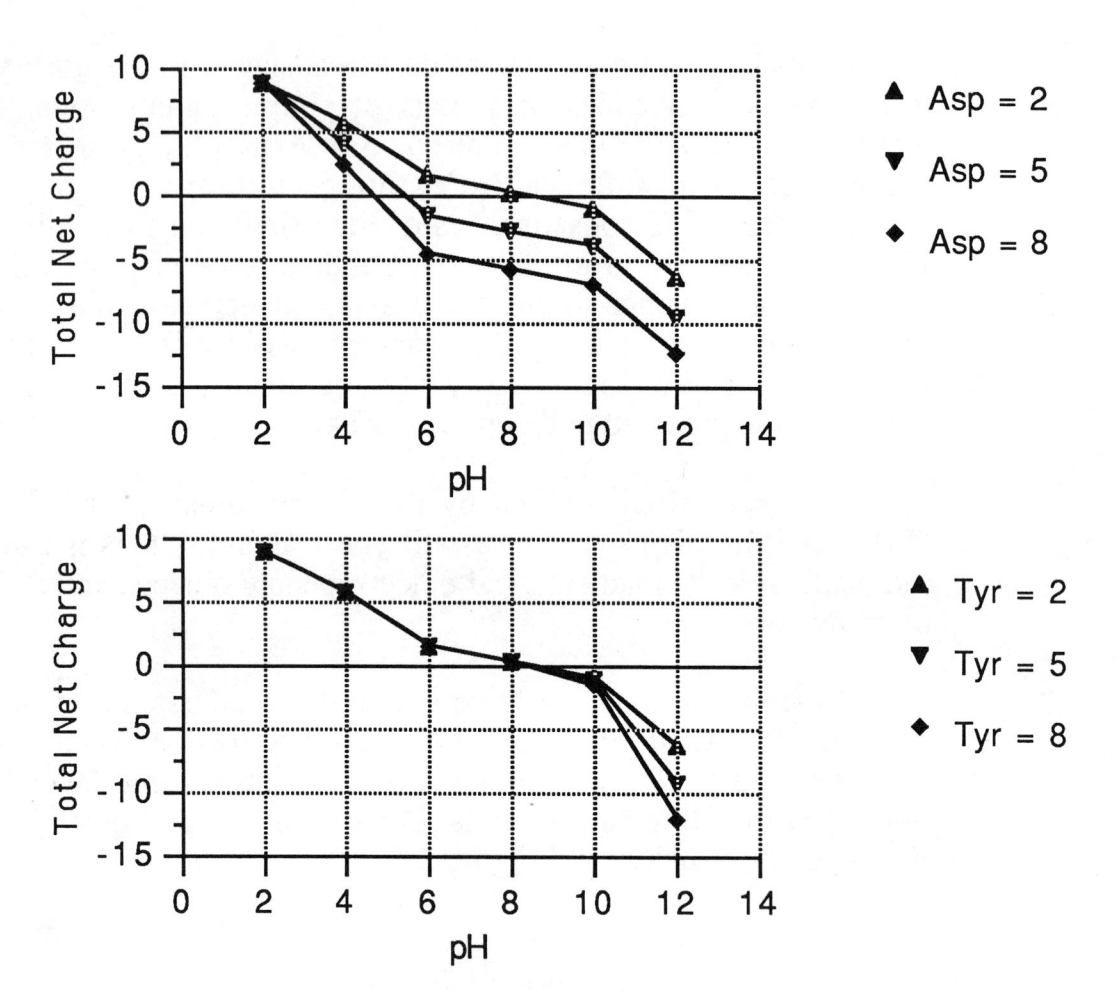

[Graph 1: Total Net Charge vs pH for Asp = 2, Asp = 5, Asp = 8]

[Graph 2: Total Net Charge vs pH for Tyr = 2, Tyr = 5, Tyr = 8]

a. As the number of aspartate residues increase in number, what happens to the shape of the pH-response curve? At what values of pH does the shape of the curve remain constant? At what values of pH does the shape of the curve change? How do these two regions relate to the pK for aspartate?

b. As the number of tyrosine residues increase in number, what happens to the shape of the pH-response curve? At what values of pH does the shape of the curve remain constant? At what values of pH does the shape of the curve change? How do these two regions relate to the pK for tyrosine?

c. What do you think would happen if you generated equivalent curves for the number of cysteine residues?

Building a Complex Model

2. Of course, computer models are capable of mimicking fantasy as well as reality. Sometimes its useful to ask "what if " of structures that have not yet been discovered. Explore the consequences of inserting an imaginary amino acid into a polypeptide chain. Make it as exotic as you wish (e.g., multiple ionizing groups on the side chain or chains).

Experiment 2.5

Forward References

You now have most of the tools you need to build the models in this book. Electronic spreadsheets have numerous additional features, but you will quickly pick them up on your own as your needs or interests dictate.

There is, however, a final caution. Some electronic spreadsheet programs are susceptible to "forward referencing errors." Even if your particular program is able to circumvent such errors, you will want to know what they are. You may even find them useful.

When a new value is entered into a cell and the entire spreadsheet is recalculated, the cells are recalculated one at a time and in a specific order. Suppose that the order of recalculation is columnwise (i.e., A1 to An, B1 to Bn, C1 to Cn). Figure 2.10 displays an example where a formula in cell B2 picks up a value from cell C1 before cell C1 is updated in the current recalculation round. If the value in cell A1 is changed from 2 to 4, cell C1 won't change from 2 * 2 to 2 * 4 until after cell B2 has been recalculated as 4 + 4 using the "wrong" value (2 * 2) from cell C3.

If the order of recalculation is rowwise, however, this problem does not exist.

If the recalculation order is columnwise, cell B1 is said to have made a *forward reference*. Forward references occur when a formula variable points forward to a cell in a region of the electronic spreadsheet that will not be recalculated until after the formula variable has already assumed a value.

If the recalculation order is columnwise (top to bottom, left to right), forward references are references to any cell to the right of the formula under consideration. If the recalculation order is rowwise (left to right, top to bottom), forward references are references to any cell below the formula under consideration.

In the example of Figure 2.10, the forward reference "clears" by simply forcing a second round of recalculation without changing any values.

Forward References

Figure 2.10

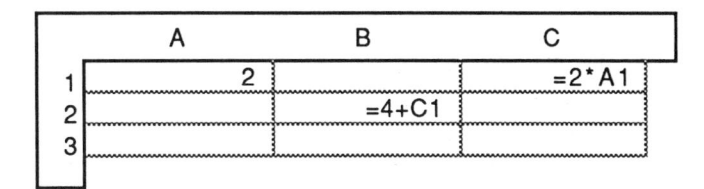

	A	B	C
1	2		=2*A1
2		=4+C1	
3			

Recalculation Order = COLUMNWISE

Forward references occur when a formula variable points forward to a cell in a region of the electronic spreadsheet that will not be calculated until after the formula under consideration has been calculated.

In the example above if the order of recalculation is columnwise (A1, A2, A3, B1, B2, B3, C1, C2, C3), cell B2 contains a forward reference to cell C1. If the order of recalculation is by row (A1, B1, C1, A2, B2, C2, A3, B3, C3), there is no forward reference.

Multiple forward references within a single model can lead to situations that require three or more recalculation rounds. Forward references can also be made to cells that reference the cell that is doing the forward references. Forward references, circular references, and self-references are all part of the joy of model building with electronic spreadsheets. (Forward references to cells containing values rather than formulas present no problems — values don't change during recalculation.)

Some newer electronic spreadsheet programs are capable of untangling forward references. When a cell is reached that contains a forward reference, the program jumps to the referenced cell, calculates the formula in that cell (if required), and returns to the original cell.

The original cell now can be evaluated without difficulty. Forward references to cells containing forward references are untangled by applying the procedure recursively. The presence of circular references can also be flagged.

The features of electronic spreadsheet programs that can untangle forward references can be turned off, and the recalculation order specified, if desired.

In this workbook, models with potential forward references will have their preferred order of recalculation specified (as in Figure 2.11).

Chapter 2   Experiment 2.5

Figure 2.11

Formulas

	A	B	C
1	2		=2*A1
2		=4+C1	
3			

Values

	A	B	C
1	2		4
2		8	
3			

Recalculation Order = ROWWISE

Directions

1. Set the order of recalculation of your electronic spreadsheet program to rowwise. Review your program instruction manual for directions if necessary.

2. Enter the model described in Figure 2.11 into your computer and verify model correctness by comparing it with Figure 2.11 (Values).

3. Save the model to a data diskette, if you wish.

4. Answer the questions at the end of the exercise.

Questions

1. Change the recalculation order of your electronic spreadsheet to columnwise. If you enter the value 4 in cell A1, what happens to the result in cell B2? What if you recalculate the spreadsheet a second time?

2. Explore the consequences of changing cell A1 to read "=C1".

3. Try to predict the behavior of the model in Figure 2.12 before you build it. Then build it and test your prediction. What happens every time you force recalculation by writing to another cell or issuing a recalculation command?

Figure 2.12

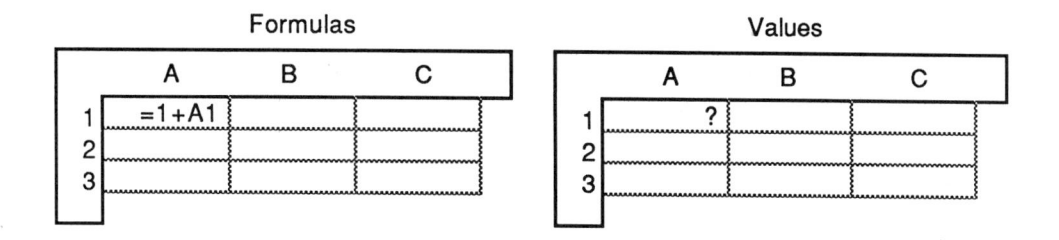

Formulas				Values		
	A	B	C		A	B	C
1	=1+A1			1	?		
2				2			
3				3			

Conclusion

An electronic spreadsheet model is an arrangement of labels, values, and formulas that are placed in an array of cells by pointing to cells with the on-screen cursor and entering cell contents from the computer keyboard.

Caution must be exercised in model design because of differences in the way different electronic spreadsheet programs

1. Identify whether a cell is a label, value or formula (e.g., B1*B2 vs. +B1*B2 vs. =B1*B2)

2. Identify and define built-in functions (e.g., LOG10(B1/B2) vs. @LOG(B1/B2))

3. Determine the order in which calculations should proceed (e.g., left-to-right vs. operator precedence rules)

4. Deal with the problem of forward references

Electronic spreadsheet programs also contain various aids that help speed the process of model design and construction. For example,

1. Commands that copy formulas from one or several cells to other cells and permit the names of formula variables to be copied without change or to change "relative" to the location of the new copy (see Experiment 2.3)

2. Editors that permit minor changes to the contents of a cell without rewriting the entire cell

Chapter 2   Experiment 2.5

3. Commands that format numeric displays as integers, fixed point, or scientific notation

4. Commands that print models to line printers or save models to storage diskettes

Your program instruction manual contains information on each of these features, and you will want to learn them all.

The construction of electronic spreadsheet models remains, however, the cycle of pointing to a cell, writing to a cell, and repeating on a new cell. If you remember this, you should have no trouble building the models and conducting the experiments in the following pages.

Chapter 3

Equilibrium and Acid-Base Relationships

Simple equilibrium relationships may seem unrelated to biochemistry. Living systems are necessarily far from equilibrium — equilibrium is death.

Equilibrium considerations, however, determine the direction of a chemical reaction. Among the most important characteristics of metabolism is the complex system of relationships that governs the direction of flow in metabolic pathways and governs the adaptability of a pathway to changes in substrate concentrations and other conditions. Equilibrium considerations are central to understanding these relationships.

Experiments in later chapters will relate the concepts of equilibrium to some aspects of metabolism. Here we merely review simple equilibrium relationships.

Experiment 3.1

Simple Equilibrium Relationships

Consider a simple reaction of the type:

$$A + B \underset{}{\overset{K_{eq}}{\rightleftharpoons}} C + D$$

The equilibrium constant K_{eq} has the form:

$$K_{eq} = \frac{[C]_{eq} [D]_{eq}}{[A]_{eq} [B]_{eq}} \qquad (3.1)$$

Chapter 3   Experiment 3.1

$[A]_{eq}$, $[B]_{eq}$, $[C]_{eq}$, and $[D]_{eq}$ are the concentrations of A, B, C, and D at equilibrium.

If the initial concentrations of A, B, C, and D are designated by $[A]_o$, $[B]_o$, $[C]_o$, and $[D]_o$, and the change in concentration required to reach equilibrium from these initial concentrations is designated by x, then the equilibrium concentrations may be written in terms of the initial concentrations:

$$[A]_{eq} = ([A]_o - x) \tag{3.2}$$
$$[B]_{eq} = ([B]_o - x) \tag{3.3}$$
$$[C]_{eq} = ([C]_o + x) \tag{3.4}$$
$$[D]_{eq} = ([D]_o + x) \tag{3.5}$$

Substituting Equations 3.2 through 3.5 into Equation 3.1 leads to the quadratic equation:

$$(1 - K_{eq})x^2 + ([C]_o + [D]_o + K_{eq}[A]_o + K_{eq}[B]_o)x + [C]_o[D]_o - K_{eq}[A]_o[B]_o = 0 \tag{3.6}$$

If Equation 3.6 is simplified by letting:

$$a = 1 - K_{eq}$$
$$b = [C]_o + [D]_o + K_{eq}[A]_o + K_{eq}[B]_o$$
$$c = [C]_o[D]_o - K_{eq}[A]_o[B]_o$$

Then:

$$x = \frac{-b + \sqrt{b^2 - 4ac}}{2a} \tag{3.7a}$$

$$x = \frac{-b - \sqrt{b^2 - 4ac}}{2a} \tag{3.7b}$$

Only Equation 3.7a has physical meaning in equilibrium chemistry.

From Equation 3.7a the fraction of A that is converted to product may be calculated as $x/[A]_o$.

Simple Equilibrium Relationships

Directions

Build and verify the model described in Figure 3.1.

Figure 3.1

Formulas

	A	B	C	D	E
1	Keq =	0.01			
2	[A]o =	1		[A]eq =	=B2-B11
3	[B]o =	2		[B]eq =	=B3-B11
4	[C]o =	0.05		[C]eq =	=B4+B11
5	[D]o =	0.02		[D]eq =	=B5+B11
6					
7	a =	=1-B1			
8	b =	=B4+B5+(B1*B2)+(B1*B3)			
9	c =	=(B4*B5)-(B1*B2*B3)			
10					
11	x =	=(-B8+SQRT(B8*B8-(4*B7*B9)))/(2*B7)			
12	x/[A]o =	=B11/B2			

Values

	A	B	C	D	E
1	Keq =	0.01			
2	[A]o =	1		[A]eq =	0.9031
3	[B]o =	2		[B]eq =	1.9031
4	[C]o =	0.05		[C]eq =	0.1469
5	[D]o =	0.02		[D]eq =	0.1169
6					
7	a =	0.9900			
8	b =	0.1000			
9	c =	-0.0190			
10					
11	x =	0.0969			
12	x/[A]o =	0.0969			

Recalculation Order = COLUMNWISE

Chapter 3   Experiment 3.1

Questions

1. Your cousin Joe works as a laboratory technician for a local chemical company. He has been assigned the task of synthesizing Superzap, a critical component in the manufacture of a profitable hygiene product. Superzap is synthesized from Zap and Reagent B via the reaction:

$$\text{Zap} + \text{Reagent B} \xrightleftharpoons{K_{eq}} \text{Superzap} + \text{Reagent D}$$

Zap is much more expensive than Reagent B.

The equilibrium constant for the reaction is small ($K = 0.01$), and so the yield of Superzap in Joe's first run is very low. He intends to overcome this problem by increasing the concentration of Zap and doesn't believe you when you tell him he's only going to make matters worse.

Use the model in Figure 3.1 to complete the table below and show your cousin his error. Why must increases in the concentration of Zap always cause the fraction of Zap converted to Superzap to decrease?

$[\text{Zap}]_0$	$[\text{B}]_0$	$[\text{Superzap}]_0$	$[\text{D}]_0$	Fraction of Zap Converted ($x/[A]_0$)
1.00	1.00	0.00	0.00	.091
25.00	1.00	0.00	0.00	.015
50.00	1.00	0.00	0.00	.010
75.00	1.00	0.00	0.00	.007
100.00	1.00	0.00	0.00	.006

When you tell Joe that he can solve his problem by increasing the concentration of Reagent B, he is delighted (Reagent B is less expensive than Zap) but skeptical. Complete the table below to show him how to obtain a higher yield of Superzap from Zap. Why do increases in [B] generate higher yields of Superzap?

Simple Equilibrium Relationships

[Zap]₀	[B]₀	[Superzap]₀	[D]₀	Fraction of Zap Converted (x/[A]₀)
1.00	1.00	0.00	0.00	.09
1.00	25.00	0.00	0.00	.38 .49
1.00	50.00	0.00	0.00	.56
1.00	75.00	0.00	0.00	
1.00	100.00	0.00	0.00	.61

2. It is always possible to force a greater conversion of one reactant by increasing the concentration of the other. The magnitude of this effect, however, depends on K_{eq}. Complete and study the table below. As [B]₀ increases 100-fold, how does the extent of increase in the fraction of Zap converted depend on the value of K_{eq}? Why does the value of K_{eq} have this effect?

Fraction of Zap Converted
([Zap]₀ = 1.00, [Superzap]₀ = [D]₀ = 0)

[B]₀	K_{eq} = 0.0001	0.001	0.01	0.1
1.00	.01	.03	.09	.24
25.00				
50.00				
75.00				
100.00	.095	.27	.61	.91

3. Suppose that either of the products Superzap or Reagent D is present at the beginning of the synthesis. Explore the effect of the ratio of [B]₀ to [D]₀ on the extent of the conversion of Zap to Superzap by completing the table below.

[Zap]₀	[B]₀	[Superzap]₀	[D]₀	Fraction of Zap Converted (x/[A]₀)
50.00	50.00	0.00	1.00	.08
50.00	20.00	0.00	1.00	.04
50.00	1.00	0.00	1.00	.005
50.00	1.00	0.00	20.00	.0004
50.00	1.00	0.00	50.00	.0002

If you wish to inhibit the conversion of Zap to Superzap would you increase or decrease the ratio of [B]₀ to [D]₀?

Chapter 3   Experiment 3.1

Problems

1. Consider the following three reactions:

$$\text{Glucose} + P_i \underset{}{\overset{K_1}{\rightleftharpoons}} \text{Glucose-6-P} + H_2O \qquad (I)$$

$$\text{Glucose} + \text{Ribose-5-P} \underset{}{\overset{K_2}{\rightleftharpoons}} \text{Glucose-6-P} + \text{Ribose} \qquad (II)$$

$$\text{Glucose} + \text{ATP} \underset{}{\overset{K_3}{\rightleftharpoons}} \text{Glucose-6-P} + \text{ADP} \qquad (III)$$

(K_1 = 0.01, K_2 = 0.99, K_3 = 1600, P_i = Phosphate ion)

Although the equilibrium constant is fixed by nature for each *reaction*, a given *conversion* (such as glucose to glucose-6-P) may have different equilibrium constants depending upon the auxiliary reagents.

a. Use the model described in Figure 3.1 to determine, by trial and error, the minimal steady-state concentration of glucose required to maintain a 5-mM concentration of glucose-6-P if inorganic phosphate is the phosphorylating agent. If ribose-5-P is the phosphorylating agent. ATP.

(In this and in the following two problems assume that: $[P_i]$ = 10 mM, [ribose-5-P]/[ribose] = 5, and [ATP]/[ADP] = 5.)

b. If glucose is available at 0.1 mM what is the maximum concentration of glucose-6-P that can be produced by orthophosphate? Ribose-5-P? ATP in a cellular environment?

(Note: By convention, the activity of pure water is defined to be 1. Thus in Equation 3.1, strict accuracy requires that $[D]_{eq}$ equal 1 when D is H_2O. For that reason, Equations 3.5 and 3.6 should be rewritten to make $[D]_o$ a constant with a value of 1 for reactions where one of the products is H_2O.)

Simple Equilibrium Relationships

2. If the ratio of [B]$_o$ to [D]$_o$ remains constant but the absolute concentrations of [B]$_o$ and [D]$_o$ increase, will the conversion of A to C increase or decrease? Try to predict the result before using your model to complete the table below for different values of K_{eq}.

[A]$_o$	[B]$_o$	[C]$_o$	[D]$_o$	K_{eq}=____ [C]$_{eq}$	K_{eq}=____ [C]$_{eq}$	K_{eq}=____ [C]$_{eq}$
50.00	20.00	0.00	1.00			
50.00	40.00	0.00	2.00			
50.00	60.00	0.00	3.00			
50.00	80.00	0.00	4.00			
50.00	100.00	0.00	5.00			

Experiment 3.2

Acidic Dissociation:
The Henderson-Hasselbalch Equation

The dissociation of a weak acid is an example of a simple but very important equilibrium relationship.

$$HB \underset{}{\overset{K_{eq}}{\rightleftharpoons}} B^- + H^+$$

The equation describing this equilibrium is, of course:

$$K_{eq} = \frac{[H^+][B^-]}{[HB]} \qquad (3.8)$$

The equation can be rewritten as the Henderson-Hasselbalch equation as either:

$$pH = pK + \log \frac{[B^-]}{[HB]} \qquad (3.9)$$

or

$$\frac{[B^-]}{[HB]} = 10^{(pH - pK)} \qquad (3.10)$$

where $pH = -\log[H^+]$
$pK = -\log[K_{eq}]$

These are the equations with which you were introduced to electronic spreadsheets in Chapter 1.

Acidic Dissociation

The mole fraction of conjugate acid (HB) and conjugate base (B) may be computed from the relationships:

$$\text{Mole Fraction (HB)} = \frac{B^-}{B^- + HB^-} = \frac{B^-/HB}{(B^-/HB) + 1} \qquad (3.11)$$

$$\text{Mole Fraction (B}^-\text{)} = 1 - \text{Mole Fraction (HB)} \qquad (3.12)$$

Directions

Build and verify the model described in Figure 3.2. Use the Copy command of your electronic spreadsheet program. The model builds a table of values of pH, [B⁻]/[HB], mole fraction of conjugate base, and mole fraction of conjugate acid for each entered value of pK.

Figure 3.2

Formulas

	A	B	C	D
1	pK =	4		
2				
3	pH	[B]/[HB]	Conj. Base	Conj. Acid
4	0	=10^(A4-B1)	=B4/(1+B4)	=1-C4
5	1	=10^(A5-B1)	=B5/(1+B5)	=1-C5
6	2	=10^(A6-B1)	=B6/(1+B6)	=1-C6
16	12	=10^(A16-B1)	=B16/(1+B16)	=1-C16
17	13	=10^(A17-B1)	=B17/(1+B17)	=1-C17
18	14	=10^(A18-B1)	=B18/(1+B18)	=1-C18

(Continued.)

Chapter 3   Experiment 3.2

Figure 3.2 (Contd.)

Values

	A	B	C	D
1	pK =	4		
2				
3	pH	[B]/[HB]	Conj. Base	Conj. Acid
4	0	0.0001	0.0001	0.9999
5	1	0.001	0.0010	0.9990
6	2	0.01	0.0099	0.9901
14	12	100000000	1.0000	0.0000
15	13	1000000000	1.0000	0.0000
16	14	10000000000	1.0000	0.0000

Questions

1. It is often useful to use electronic spreadsheet programs to generate tables of values that result from simple calculations. Use the model in Figure 3.2 to complete the table below:

pK	pH	[B⁻]/[HB]	Mole Fraction Conjugate Base	Mole Fraction Conjugate Acid
4	2			
4	4			
4	6			
4	8			
4	10			
4	12			
8	2			
8	4			
8	6			
8	8			
8	10			
8	12			

When pH is less than pK is [B⁻]/[BH] greater than, less than, or equal to 1? When pH is greater than pK? When pH is equal to pK?

40

Acidic Dissociation

2. Complete the table below and graph the results. It will show you the dissociation behavior of each of the acidic groups of the amino acid arginine.

pK	pH	[B$^-$]/[HB]	Mole Fraction Conjugate Base	Mole Fraction Conjugate Acid
2.3	0	5.01 × 10^{-3}	0.00499	0.995
2.3	2	0.501	0.334	0.666
2.3	4	50.1	0.980	0.0196
2.3	6	5.01 × 10^{3}	0.9998	2.0 × 10^{-4}
2.3	8	5.01 × 10^{5}	≈1.000	2.0 × 10^{-6}
2.3	10	5.01 × 10^{7}	≈1.000	2.0 × 10^{-8}
2.3	12	5.01 × 10^{9}	≈1.000	2.0 × 10^{-10}
2.3	14	5.01 × 10^{11}	≈1.000	2.0 × 10^{-12}
9.1	0	7.94 × 10^{-10}	7.94 × 10^{-10}	≈1.000
9.1	2	7.94 × 10^{-8}	7.94 × 10^{-8}	≈1.000
9.1	4	7.94 × 10^{-6}	7.94 × 10^{-6}	≈1.000
9.1	6	7.94 × 10^{-4}	7.93 × 10^{-4}	0.999
9.1	8	0.0794	0.0736	0.926
9.1	10	7.94	0.888	0.112
9.1	12	794	0.9987	0.00126
9.1	14	7.94 × 10^{4}	≈1.000	1.26 × 10^{-5}
12.5	0	3.16 × 10^{-13}	3.16 × 10^{-13}	≈1.000
12.5	2	3.16 × 10^{-11}	3.16 × 10^{-11}	≈1.000
12.5	4	3.16 × 10^{-9}	3.16 × 10^{-9}	≈1.000
12.5	6	3.16 × 10^{-7}	3.16 × 10^{-7}	≈1.000
12.5	8	3.16 × 10^{-5}	3.16 × 10^{-5}	≈1.000
12.5	10	3.16 × 10^{-3}	0.00315	0.997
12.5	12	0.316	0.240	0.760
12.5	14	31.6	0.969	0.0307

△ pK = 2.3
▽ pK = 9.1
◇ pK = 12.5

Chapter 3   Experiment 3.2

a. The pKs of arginine's three ionizable groups have very different values, and only four of the eight imaginable ionic forms of arginine ever exist in appreciable concentrations.

Study the ionization pathway for arginine below.

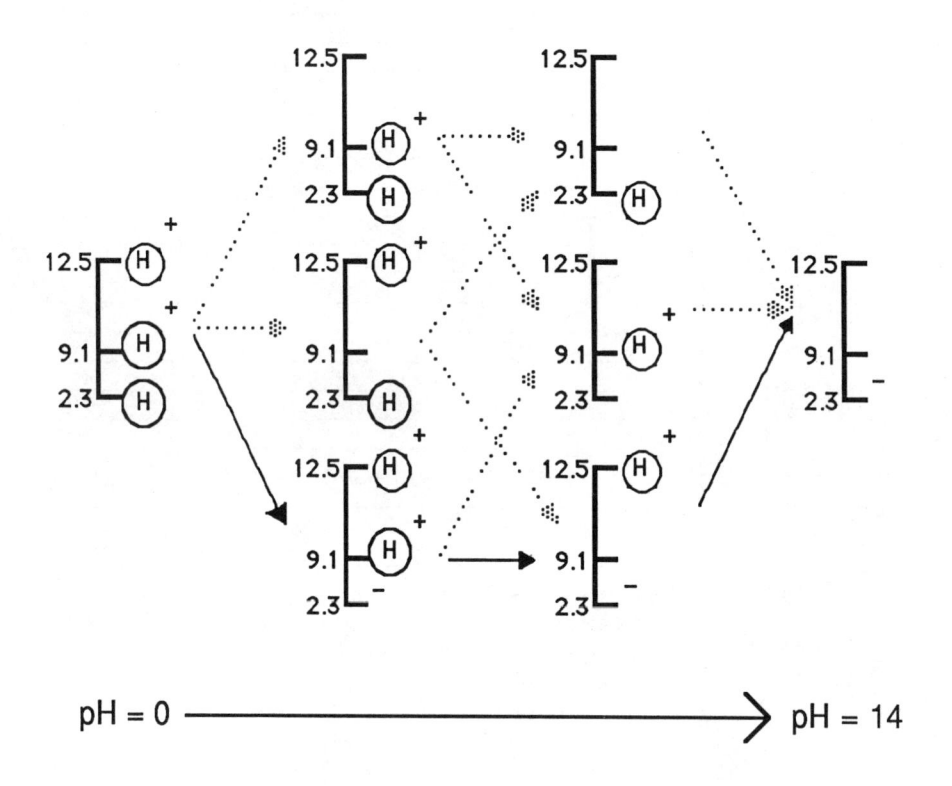

pH = 0 ⟶ pH = 14

3. Repeat the exercises in Experiment 2.4. Can you explain why changes in the number of specific residues have the effects that they do?

42

Experiment 3.3

Arginine Ionic Forms versus pH

The pKs of the three ionizing groups of arginine are widely separated. Determination of the concentrations of the ionic forms of arginine at a specific pH need only consider, therefore, four of the eight imaginable ionic forms (Figure 3.3 and Experiment 3.2, Question 2).

Figure 3.3

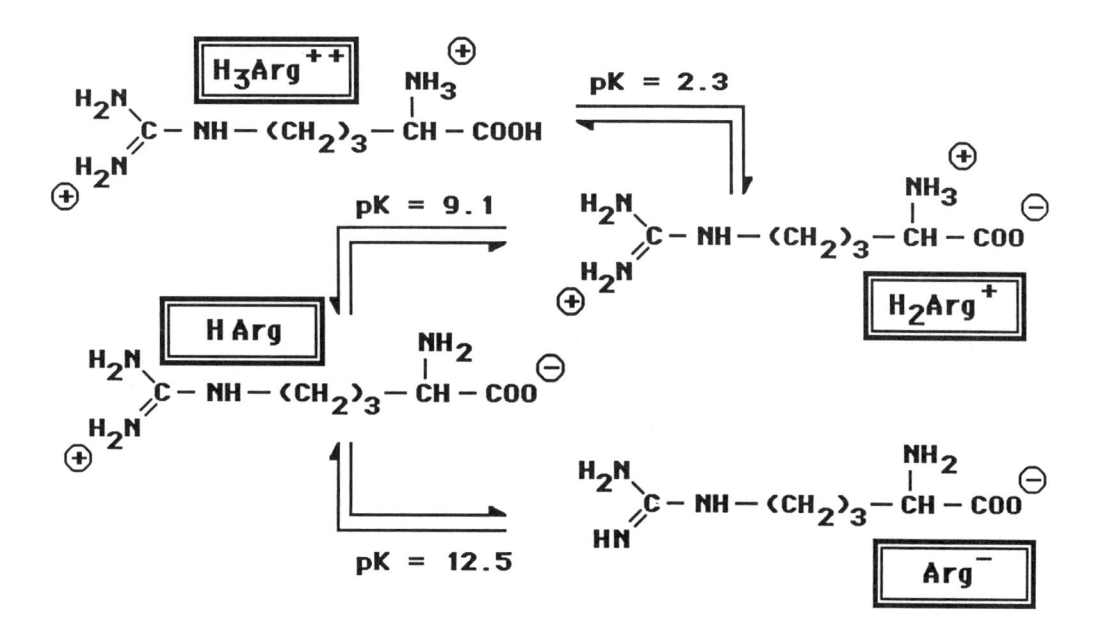

Chapter 3   Experiment 3.3

The ratios of each pair of ionic forms can be calculated for a given pH by using the Henderson-Hasselbalch equation:

$$\frac{[H_2 Arg^+]}{[H_3 Arg^{++}]} = 10^{(pH - 2.3)} \qquad (3.13)$$

$$\frac{[HArg]}{[H_2 Arg^+]} = 10^{(pH - 9.1)} \qquad (3.14)$$

$$\frac{[Arg^-]}{[HArg]} = 10^{(pH - 12.5)} \qquad (3.15)$$

$[H_3Arg^{++}]$, expressed as a percentage of the total concentration of all ionic forms, may be calculated from the ratios obtained in Equations 3.13 through 3.15 by using the partition equation:

$$100 = [H_3Arg^{++}] + [H_2Arg^+] + [HArg] + [Arg^-] \qquad (3.16)$$

or

$$100 = [H_3 Arg^{++}] \left[1 + \left(\frac{[H_2 Arg^+]}{[H_3 Arg^{++}]}\right) + \left(\frac{[HArg]}{[H_2 Arg^+]}\right)\left(\frac{[H_2 Arg^+]}{[H_3 Arg^{++}]}\right) \right.$$
$$\left. + \left(\frac{[Arg^-]}{[HArg]}\right)\left(\frac{[HArg]}{[H_2 Arg^+]}\right)\left(\frac{[H_2 Arg^+]}{[H_3 Arg^{++}]}\right) \right] \qquad (3.17)$$

Equation 3.17 is solved for $[H_3Arg^{++}]$, and the concentrations of the other ionic forms are calculated from $[H_3Arg^{++}]$ and the previously computed concentration ratios (Equations 3.13 through 3.15):

$$[H_2 Arg^+] = \frac{[H_2 Arg^+]}{[H_3 Arg^{++}]} [H_3 Arg^{++}] \qquad (3.18)$$

$$[HArg] = \frac{[HArg]}{[H_2 Arg^+]} [H_2 Arg^+] \qquad (3.19)$$

$$[Arg^-] = \frac{[Arg^-]}{[HArg]} [HArg] \qquad (3.20)$$

Arginine Ionic Forms

Thus the concentrations of each of the principal ionic forms of arginine may be calculated for any given pH.

Partition equations will be used frequently in this workbook. Such treatments form a recurring theme in biochemistry.

Directions

Build and verify the model described in Figure 3.4.

Figure 3.4

Formulas

	A	B	C	D	E
1	pK1 =	2.3			
2	pK2 =	9.6			
3	pK3 =	12.5			
4	pH =	7.2			
5					
6	[H2Arg]/[H3Arg] =	=10^(B4-B1)			
7	[HArg]/[H2Arg] =	=10^(B4-B2)			
8	[Arg]/[HArg] =	=10^(B4-B3)			
9					
10	100/[H3Arg] =	=1+B6+(B7*B6)+(B8*B7*B6)			
11					
12	Ionic Form	Chrg/Molec	Mole Frctn	Fractional Charge	
13	H3Arg	2	=1/B10	=B13*C13	
14	H2Arg	1	=B6*C13	=B14*C14	
15	HArg	0	=B7*C14	=B15*C15	
16	Arg	-1	=B8*C15	=B16*C16	
17				=SUM(D13..D16)	= Tot. Chrg.

Values

	A	B	C	D	E
1	pK1 =	2.3			
2	pK2 =	9.6			
3	pK3 =	12.5			
4	pH =	7.2			
5					
6	[H2Arg]/[H3Arg] =	79432.8235			
7	[HArg]/[H2Arg] =	0.0040			
8	[Arg]/[HArg] =	5.01E-6			
9					
10	100/[H3Arg] =	79750.0528			
11					
12	Ionic Form	Chrg/Molec	Mole Frctn	Fractional Charge	
13	H3Arg	2	1.25E-5	2.51E-5	
14	H2Arg	1	0.9960	0.9960	
15	HArg	0	3.97E-3	0	
16	Arg	-1	1.99E-8	-1.99E-8	
17				0.9960	= Tot. Chrg.

Chapter 3 Experiment 3.3

Questions

1. Determine the concentration (mole fraction) of each ionic form of arginine over a range of pHs by completing and graphing the table below:

pH	[H$_3$Arg^{++}]	[H$_2$Arg$^+$]	[HArg]	[Arg$^-$]
1	____	____	____	____
2	____	____	____	____
3	____	____	____	____
4	____	____	____	____
5	____	____	____	____
6	____	____	____	____
7	____	____	____	____
8	____	____	____	____
9	____	____	____	____
10	____	____	____	____
11	____	____	____	____
12	____	____	____	____
13	____	____	____	____
14	____	____	____	____

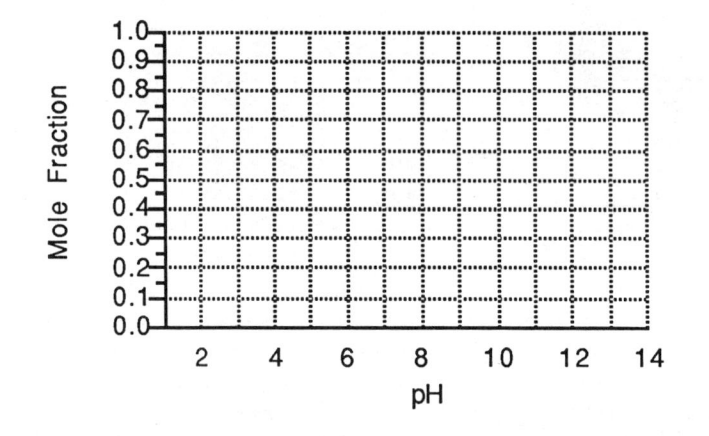

Graph: Mole Fraction (0.0 to 1.0) vs. pH (2 to 14)

- △ [H$_3$Arg^{++}]
- ▽ [H$_2$Arg$^+$]
- □ [HArg]
- ○ [Arg$^-$]

2. By trial and error, find the isoelectric point, the pH at which the sum of the electrical charges of all arginine ionic forms equals zero. Which ionic form predominates at this pH, and what is its net electrical charge? What is the predominant form at physiological pH (pH = 7.2), and what is its net electrical charge?

3. Plot the net electrical charge, summed over all four significant arginine ionic forms, as a function of pH.

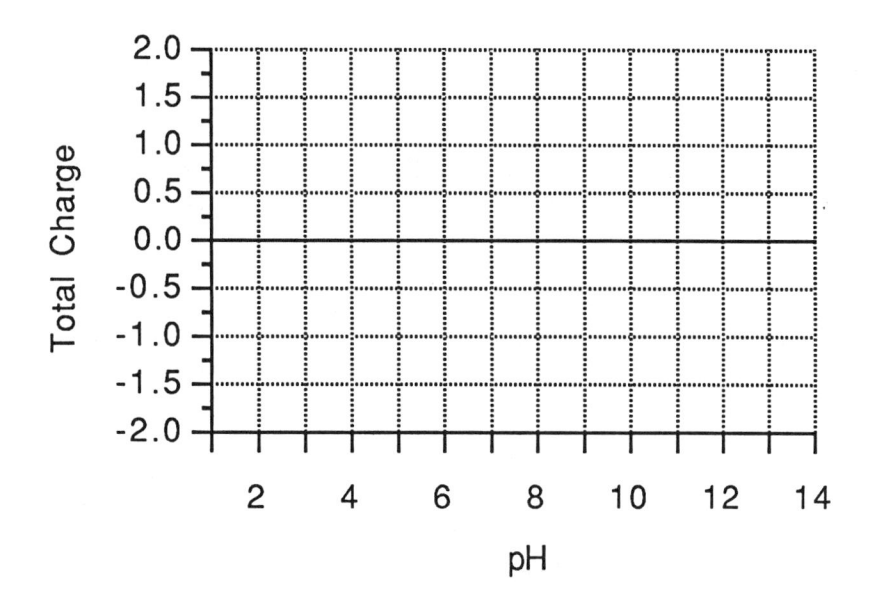

a. Identify the pH values on the bottom axis of the graph that are numerically equal to the pKs of arginine.

b. Compare the shape of this curve with the curves that you plotted in Question 1. Note the correspondence, for example, of the pH range during which the total charge is approximately +1 and the pH range during which H_2Arg^+ is the predominant ionic form of arginine. What other correspondences do you see?

c. What is the relationship between the graph above and a standard titration curve?

Problems

1. Modify the model described in Figure 3.4 so that it describes the behavior of lysine ($pK_1 = 2.2$, $pK_2 = 9.2$, $pK_3 = 10.8$; net charge of the fully protonated form is +2). Find the isoelectric point and the predominant ionic form at that pH. What is the predominant ionic form at pH = 7.2?

Chapter 3   Experiment 3.3

2. What is the isoelectric point of histidine? What is its predominant ionic form at the isoelectric point? At pH 7.2? (For histidine, pK_1 = 1.8, pK_2 = 6.0, pK_3 = 9.2; net charge of the fully protonated form is +2.)

3. What is the isoelectric point of tyrosine? What is its predominant ionic form at the isoelectric point? At pH 7.2? (For tyrosine, pK_1 = 2.2, pK_2 = 9.1, pK_3 = 10.9; net charge of the fully protonated form is +1.)

4. The model described in Figure 3.4 assumes that the pKs of the ionizing groups are widely separated. Thus, the model follows the main sequence of ionizations and ignores the other imaginable ionic forms. Extend the model to include all imaginable ionic forms. At pH = 2, 7, and 10, what percent of the total arginine is present in these minor forms? Compare these results with the situation where two of the ionization constants are relatively close together (e.g., cysteine: pK_1 = 1.8, pK_2 = 8.3, pK_3 = 10.8).

5. This model is a specific example of a general case that will be developed in Experiment 5.2, "Multiple Sequential Equilibria." You may want to skip ahead to that experiment and solidify your knowledge of this topic. Notice also that, although developed differently, the model described in Figure 3.4 is very similar to that described in Experiment 2.4 (Figure 2.6).

Experiment 3.4

Buffers

Although buffers are first encountered in beginning chemistry classes, even many advanced students don't fully understand the chemical basis of buffer action. Buffers are taken for granted and often are chosen inappropriately through lack of thought.

When strong acid is added to a buffered system the concentration of conjugate base decreases, and the concentration of conjugate acid increases by the same amount. Addition of strong base forces the reaction in the opposite direction as shown.

$$\text{(Base added)} \\ HB \rightleftharpoons H^+ + B^- \\ \text{(Acid added)}$$

If the concentration of buffer (conjugate acid plus conjugate base) is assigned a value of 1, the Henderson-Hasselbalch equation may be written:

$$pH = pK + \log \frac{[B^-]}{[1-B^-]} \qquad (3.21)$$

A buffer is most effective at its midpoint when the concentrations of the conjugate acid and conjugate base are equal (i.e., when pH = pK). This is a consequence of the properties of a ratio of the form $x/(1-x)$. A change in the value of x has the least effect on the value of this ratio when $x = (1-x) = 0.5$ (Figure 3.5a). This also can be seen clearly in a plot of $\log[x/(1-x)]$ vs. x (Figure 3.5b).

Chapter 3   Experiment 3.4

Figure 3.5

$$f(x) = \frac{x}{1-x}$$

(a)

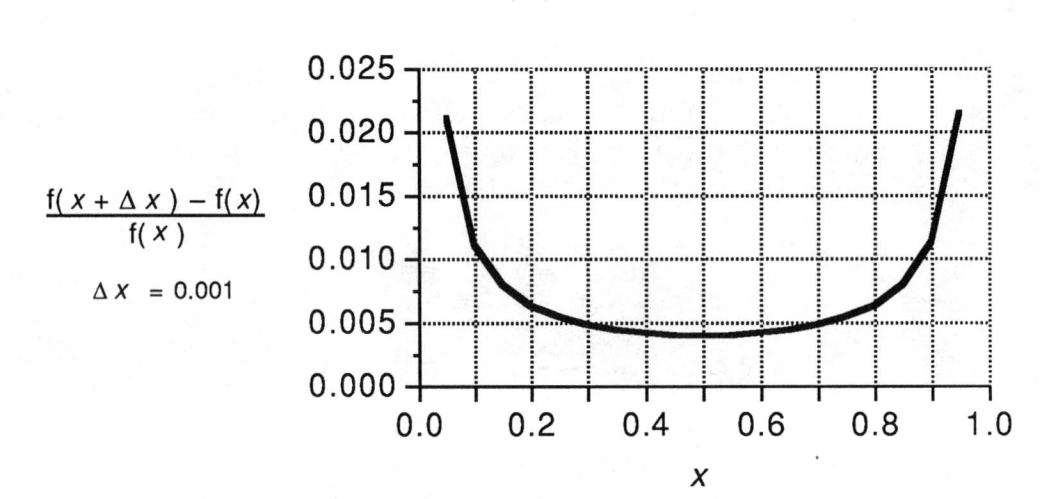

$\dfrac{f(x+\Delta x) - f(x)}{f(x)}$

$\Delta x = 0.001$

(b)

log f(x)

50

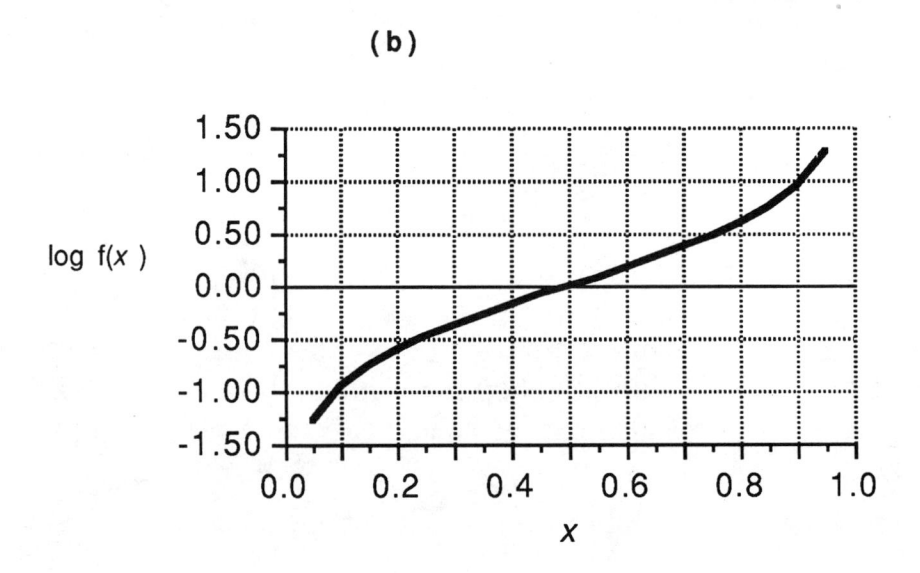

Buffers

Directions

1. Build and verify the model described in Figure 3.6.

2. The model generates a table of [B⁻]/[HB], pH, and ΔpH/Δ[B⁻] (a measure of buffering capacity) for a user-defined range of [B⁻] and [HB] values. The content of cell B2 defines the beginning of the table; cell B3 defines the increment between each table value. For most purposes, the total buffer present (cell E1) can be conveniently left at 100.

3. Notice that several cells display "ERROR" for values. ERROR simply means that the formula in that cell is mathematically unacceptable (e.g., $\log_{10}$ of 0) or that a formula variable points to a cell that is mathematically unacceptable.

Figure 3.6

Formulas

	A	B	C	D	E
1	pKa =	7.2		[B] + [HB] =	100
2	Start [B] =	0			
3	Incr [B] =	10			
4					
5	[B]	[HB]	[B]/[HB]	pH	ΔpH/Δ[B]
6	=B2	=E1-A6	=A6/B6	=B1+LOG(C6)	
7	=A6+B3	=E1-A7	=A7/B7	=B1+LOG(C7)	=D7-D6/B3
8	=A7+B3	=E1-A8	=A8/B8	=B1+LOG(C8)	=D8-D7/B3
9	=A8+B3	=E1-A9	=A9/B9	=B1+LOG(C9)	=D9-D8/B3

Values

	A	B	C	D	E
1	pKa =	7.2		[B] + [HB] =	100
2	Start [B] =	0			
3	Incr [B] =	10			
4					
5	[B]	[HB]	[B]/[HB]	pH	ΔpH/Δ[B]
6	0	100	0.0000	ERR	
7	10	90	0.1111	6.2458	ERR
8	20	80	0.2500	6.5979	0.0352
9	30	70	0.4286	6.8320	0.0234

Chapter 3   Experiment 3.4

Questions

1. Suppose that you add 10 mmole (mmol) increments of strong acid to 1 L of a 100 mM buffer solution (pK = 6.8) that contains 10 mM conjugate acid (and therefore 90 mM conjugate base). Assume that the volume changes are negligible. Build the table below by using the model described in Figure 3.6 and determine (a) the pH of the freshly prepared buffer and (b) the pH of the buffer at each additional increment of acid. Graph your results. (**Note:** Add acid to your model by incrementing [B] (cell B3) with a negative value.)

[B⁻]	[HB]	pH
90	10	___
80	20	___
70	30	___
60	40	___
50	50	___
40	60	___
30	70	___
20	80	___
10	90	___

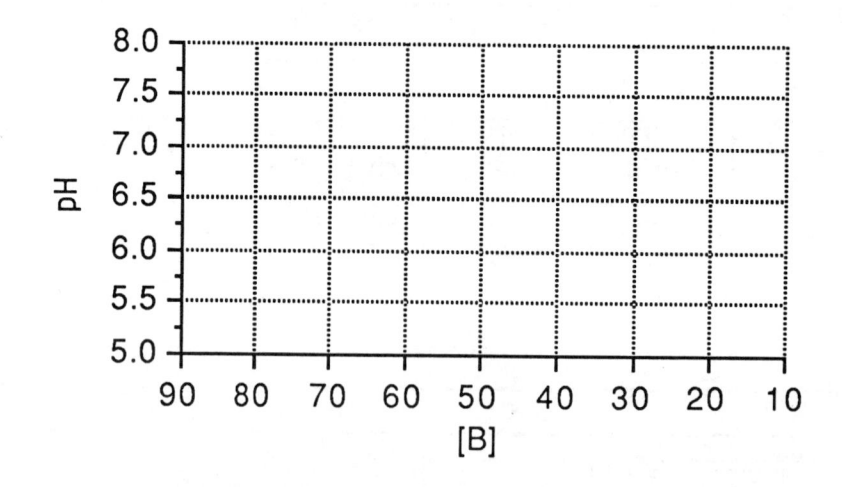

2. As strong acid is added to the buffer described in Question 1, the effectiveness of the buffer changes. Complete and graph the table below to determine the change in pH for each 10 mM change in [B] (i.e., for each addition of 10 mmol liter^{-1} of acid).

[B⁻]	[HB]	ΔpH/Δ[B⁻]
90	10	_____
80	20	_____
70	30	_____
60	40	_____
50	50	_____
40	60	_____
30	70	_____
20	80	_____
10	90	_____

3. In Question 2, buffer effectiveness was measured as the change in pH vs. a 10-mM change in [B⁻]. As the change in [B⁻] approaches zero, ΔpH/Δ[B] approaches the *instantaneous* rate of change of pH with respect to [B⁻].

What is the instantaneous rate of change of pH with respect to added base (the rate of change of pH as the rate of change of added base approaches zero) when the buffer described in Question 2 contains 40 mM [B⁻] and 60 mM [HB]?

Chapter 3 Experiment 3.4

[B⁻]	[HB]	pH	ΔpH/Δ[B⁻]*
40	60	_____	
45	55	_____	_____
41	59	_____	_____
40.1	59.9	_____	_____
40.01	59.99	_____	_____

*All changes measured against [B⁻] = 40, [HB] = 60 with changes made in "Incr [B⁻]"

4. What is the instantaneous rate of change of pH with respect to added acid when the buffer described in Question 2 contains 70-mM [B] and 30-mM [HB]?

5. For the buffer described in Question 1, verify that the lowest instantaneous rate of change of pH with respect to added acid occurs when pH = pK.

Problems

1. It is sometimes said that at a distance of 1 pH unit from the midpoint of a buffer (i.e., where the ratio [B⁻]/[HB] is either 10 or 0.1), a buffer is only one-tenth as effective (measured as the instantaneous rate of change of pH with with respect to [B⁻]) as at the midpoint. Is this true? How far from the midpoint (in terms of units of pH) must you travel before the slope is double that at the midpoint?

2. A buffer is most effective at its midpoint, but it is not always advantageous to begin an experiment at the midpoint pH. Suppose that you wish to assay an enzyme at pH 7.2 and have luckily found a buffer with pK equal to 7.2. However, the enzyme is poisoned by buffer at high concentrations, and you cannot use a concentration greater than 100 mM. Suppose that the enzyme reaction will release 30 mmoles of proton/liter, and you wish to hold the pH as close to pH = 7.2 as possible. Should the reaction begin at pH 7.2 or at some other value? If some other value, what should that value be if the buffer concentration is 100 mM?

Experiment 3.5

Partition Between Phases as a Function of pH

Consider a weak acid distributed on the basis of polarity across the boundary separating an organic and an aqueous phase (Figure 3.7).

The principles that determine this distribution govern the behavior of many important laboratory procedures in biochemistry and influence the distribution of molecules among the compartments of living cells.

Figure 3.7 might describe the behavior of a carboxylic acid: The conjugate acid is uncharged, and the conjugate base bears a single negative charge. The conjugate acid, being uncharged and relatively nonpolar, is more soluble in the organic phase than in water. The conjugate base is very much more polar because of its negative charge and is therefore less soluble in the organic phase, and more soluble in water, than the conjugate acid.

Figure 3.7

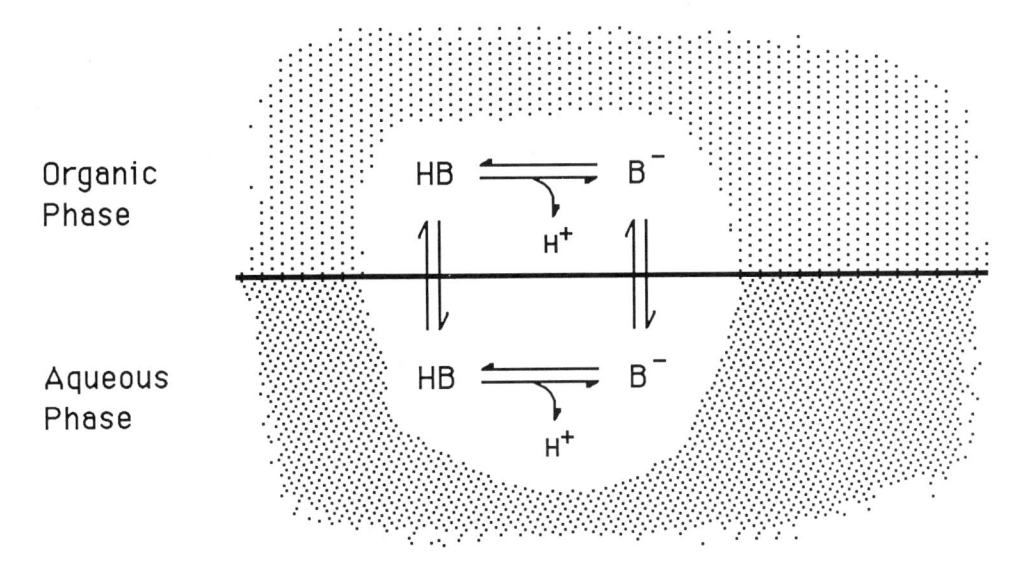

Chapter 3   Experiment 3.5

If the total amount of conjugate acid and conjugate base to be distributed between the two phases is assumed to be 100 mmoles (and, for convenience, the volume of each phase is assumed to be 1 L), then:

$$100 = [HB]_w + [HB]_o + [B]_w + [B]_o \quad (3.22)$$

where $[HB]_w$ = Concentration of conjugate acid in the aqueous phase
    $[HB]_o$ = Concentration of conjugate acid in the organic phase
    $[B]_w$ = Concentration of conjugate base in the aqueous phase
    $[B]_o$ = Concentration of conjugate base in the organic phase

The partition equation may be rearranged in a fashion similar that displayed in Experiment 3.3 to yield:

$$100 = [HB]_w \left[1 + \frac{[HB]_o}{[HB]_w} + \frac{[B]_w}{[HB]_w}\left(1 + \frac{[B]_o}{[B]_w}\right) \right] \quad (3.23)$$

Because $[B]_w/[HB]_w$ can be computed using the Henderson-Hasselbalch equation and $[HB]_o/[HB]_w$ and $[B]_o/[B]_w$ are determined by the solubilities of each component in the organic and aqueous phases, Equation 3.23 is written:

$$100 = [HB]_w \left[1 + \frac{S[HB]_o}{S[HB]_w} + 10^{(pH-pK)}\left(1 + \frac{S[B]_o}{S[B]_w}\right) \right] \quad (3.24)$$

where $S(HB_o)$ = Solubility of conjugate acid in the organic phase
    $S(HB_w)$ = Solubility of conjugate acid in the aqueous phase
    $S(B_o)$ = Solubility of conjugate base in the organic phase
    $S(B_w)$ = Solubility of conjugate base in the aqueous phase

Equation 3.24 may be solved for $[HB]_w$ and the remaining terms of Equation 3.22 solved using the relationships:

$$[HB]_o = [HB]_w \frac{S[HB]_o}{S[HB]_w} \quad (3.25)$$

$$[B]_w = \frac{[B]_w}{[HB]_w}[HB]_w = 10^{(pH-pK)}[HB]_w \quad (3.26)$$

$$[B]_o = [B]_w \frac{S[B]_o}{S[B]_w} \quad (3.27)$$

Partition Between Phases

Directions

Build and verify the model described in Figure 3.8.

Figure 3.8

Formulas

	A	B	C	D	E	F	G	H
1		Total =	100	S(HBw) =	0.1			
2		pK =	4	S(Bw) =	100			
3				S(HBo) =	10			
4				S(Bo) =	0.01			
5								
6	pH	[B]w/[HB]w	[HB]w	[B]w	[HB]o	[B]o	[Tot]w	[Tot]o
7	2.0	=10^(A7-C2)		=C7*B7	=C7*(E3/E1)	=D7*(E4/E2)	=C7+D7	=E7+F7
8	2.5	=10^(A8-C2)		=C8*B8	=C8*(E3/E1)	=D8*(E4/E2)	=C8+D8	=E8+F8
9	3.0	=10^(A9-C2)	Formula	=C9*B9	=C9*(E3/E1)	=D9*(E4/E2)	=C9+D9	=E9+F9
10	3.5	=10^(A10-C2)	Set I	=C10*B10	=C10*(E3/E1)	=D10*(E4/E2)	=C10+D10	=E10+F10
11.0		=10^(A25-C2)		=C25*B25	=C25*(E3/E1)	=D25*(E4/E2)	=C25+D25	=E25+F25
11.5		=10^(A26-C2)		=C26*B26	=C26*(E3/E1)	=D26*(E4/E2)	=C26+D26	=E26+F26
12.0		=10^(A27-C2)		=C27*B27	=C27*(E3/E1)	=D27*(E4/E2)	=C27+D27	=E27+F27

Values

	A	B	C	D	E	F	G	H
1		Total =	100	S(HBw) =	0.1			
2		pK =	4	S(Bw) =	100			
3				S(HBo) =	10			
4				S(Bo) =	0.01			
5								
6	pH	[B]w/[HB]w	[HB]w	[B]w	[HB]o	[B]o	[Tot]w	[Tot]o
7	2.0	1.00E-2	0.9900	0.0099	99.0001	0.0000	0.9999	99.0001
8	2.5	3.16E-2	0.9898	0.0313	98.9789	0.0000	1.0211	98.9789
9	3.0	1.00E-1	0.9891	0.0989	98.9120	0.0000	1.0880	98.9120
10	3.5	3.16E-1	0.9870	0.3121	98.7008	0.0000	1.2991	98.7009
11.0		1.00E7	0.0000	99.9890	0.0010	0.0100	99.9890	0.0110
11.5		3.16E7	0.0000	99.9897	0.0003	0.0100	99.9897	0.0103
12.0		1.00E8	0.0000	99.9899	0.0001	0.0100	99.9899	0.0101

Formula Set I
Prototype Cell is C7
=C1/(((1+(E3/E1))+(B7*(1+(E4/E2))))

Chapter 3   Experiment 3.5

Questions

1. Complete the table below to observe how pK influences the distribution of each component between the organic and aqueous phases. Assume that the total of both components is 100 mmol and the volume of each phase is 1 L. Graph your results.

$S(HB_w) = 0.1$ $S(B_w) = 100$
$S(HB_o) = 10$ $S(B_o) = 0.01$

pK = 4	pH	$[HB]_w$	$[B]_w$	$[HB]_o$	$[B]_o$
	3				
	5				
	7				
	9				
	11				

pK = 7	pH	$[HB]_w$	$[B]_w$	$[HB]_o$	$[B]_o$
	3				
	5				
	7				
	9				
	11				

pK = 10	pH	$[HB]_w$	$[B]_w$	$[HB]_o$	$[B]_o$
	3				
	5				
	7				
	9				
	11				

Partition Between Phases

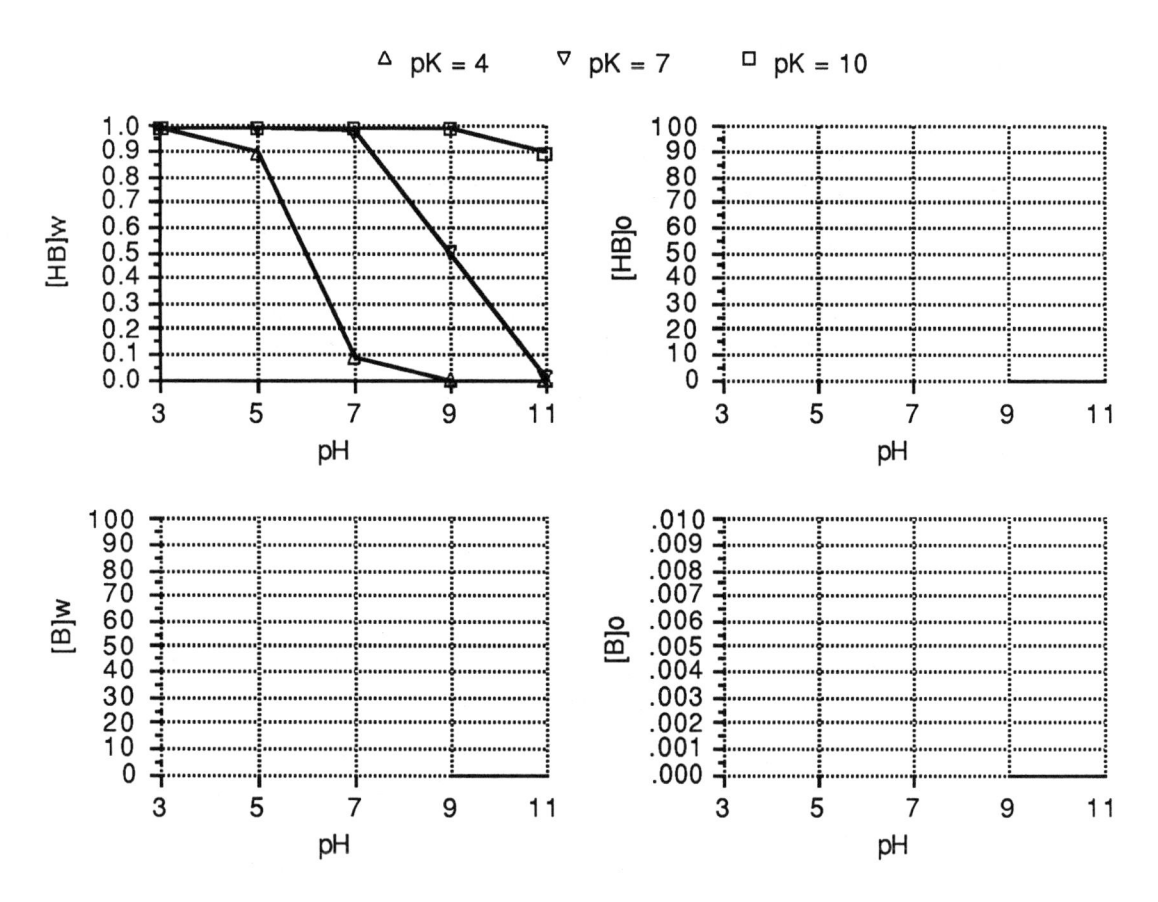

△ pK = 4   ▽ pK = 7   □ pK = 10

[Graphs: top-left shows [HB]w vs pH with data plotted for the three pK values; top-right [HB]o vs pH (empty); bottom-left [B]w vs pH (empty); bottom-right [B]o vs pH (empty).]

2. If the solubility of the conjugate acid in water increases ten-fold, how do the results in Question 1 change? What would be the effect of a five-fold decrease in the solubility of the conjugate acid in the organic phase?

3. Because the conjugate acid is much more soluble in the organic phase and the conjugate base is much more soluble in the aqueous phase, it might seem that the solute should distribute equally between the phases when pH equals pK. Is this true? Why or why not?

4. If more solute were added to the aqueous phase at low values of pH, most of it would move into the organic phase even though the concentration is already higher in that phase. How do you explain the movement of solute against a concentration gradient?

59

Chapter 3   Experiment 3.5

Problems

1. Modify the model described in Figure 3.8 such that the conjugate acid is charged and the conjugate base is uncharged (e.g., an organic amine).

2. The model described in Figure 3.8 assumes that the volumes of the two phases are equal. Modify the model so that the volumes of each of the phases may be treated as variables.

3. Naturally occurring amino acids exist almost entirely as dipolar ions at physiological pH. The concentration of the uncharged molecule is extremely low (see Experiment 3.3).

    Build a model that simulates the distribution of anthranilic acid (2-aminobenzoic acid) across the boundary between organic and aqueous phases at different values of pH. The pK values for the carboxylic group and for the amine are 4.9 and 2.1, respectively. It is sufficient to set the solubility of the uncharged molecule in the organic phase to a much higher value than its solubility in the aqueous phase. The opposite holds true for the charged molecule.

    a. How do the relative concentrations of the different ionic species of anthranilic acid vary with pH? Compare the ratio of concentrations of the dipolar ion and the uncharged molecule when pH lies between the two pKa values with the corresponding ratio for an amino acid.

    b. How does anthranilic acid distribute between water and ether as a function of pH?

    c. How might you separate alanine, anthranilic acid, benzoic acid, and aniline if you had only water, ether, and control over the pH of the aqueous phase? (For alanine, pK values are 2.3 and 9.6, for benzoic acid and aniline, the pK values are 4.2 and 4.6, respectively.)

Experiment 3.6

Distribution of Ammonia Between Blood and Urine

Consider a membrane that is permeable only to the conjugate base (or conjugate acid) of a weak acid. If the solutions on opposite sides of the membrane have different pH values, the form of the weak acid to which the membrane is not permeable will exist at different concentrations in the two regions. The distribution of ammonia between blood and urine is a much-studied and important example.

NH_3 (but not NH_4^+) equilibrates across the barrier formed by the renal tubule cells and therefore exists at the same concentration on either side of the barrier. However, the pH of the urine is usually lower than the pH of the blood because the tubule cells secrete protons into the urine. The Henderson-Hasselbalch equation shows that the concentration of NH_4^+ in the urine will be greater than in the blood (Figure 3.9).

This experiment will illustrate the unequal distribution of a weak acid or weak base across a membrane that is permeable to only one of its conjugate forms. The actual situation in the kidney is more complex and is represented incompletely by this model. For example, ammonium ion is produced in the kidney by hydrolysis of glutamine, and under most conditions the kidney loses more ammonia to the blood than it takes up. However, the principle illustrated here is valid, and the kidney is a major example of its application.

Chapter 3 Experiment 3.6

Figure 3.9

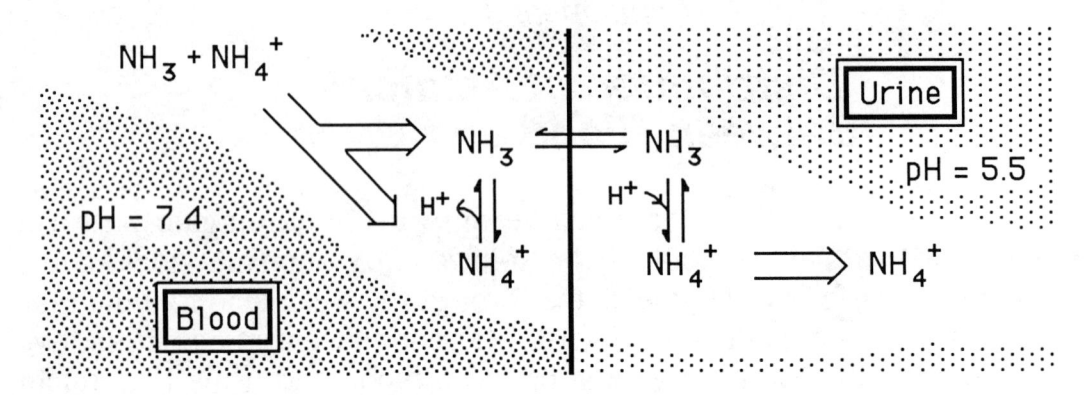

Ammonia (NH_3 and NH_4^+) is transported by the blood to the vicinity of the renal tubule where NH_3 equilibrates with urine across a barrier that is impermeable to NH_4^+. The renal cells that constitute the blood-urine barrier secrete H⁺ into the urine and appreciably reduce the pH of the urine with respect to the blood. Consideration of the Henderson-Hasselbalch equation reveals that if NH_3 concentrations on both sides of the barrier are the same, the concentration of NH_4^+ must be greater in the urine than in the blood.

In the blood, the known quantities are:

Concentration of total ammonia: $[Tot]_{Bl}$ (= $[NH_3 + NH_4^+]$)
Blood pH: pH_{Bl}
pK of the dissociable ammonia proton: pK (= 9.25)

The quantities that may be calculated are:

$$\frac{[NH_3]_{Bl}}{[NH_4^+]_{Bl}} = 10^{(pH_{Bl} - pK)} \tag{3.28}$$

$$[NH_3]_{Bl} = \frac{[NH_3]_{Bl}}{[Tot]_{Bl}}[Tot]_{Bl} = \left(\frac{[NH_3]_{Bl}}{[NH_3]_{Bl} + [NH_4^+]_{Bl}}\right)[Tot]_{Bl}$$

$$= \frac{\left(\frac{[NH_3]_{Bl}}{[NH_4^+]_{Bl}}\right)}{\left(\frac{[NH_3]_{Bl}}{[NH_4^+]_{Bl}} + 1\right)}[Tot]_{Bl} \tag{3.29}$$

$$[NH_4^+]_{Bl} = [Tot]_{Bl} - [NH_3]_{Bl} \tag{3.30}$$

Distribution of Ammonia

In the urine, if the pH is known, the quantities that may be calculated are:

$$\frac{[NH_3]_{Ur}}{[NH_4^+]_{Ur}} = 10^{(pH_{Ur} - pK)} \qquad (3.31)$$

$$[NH_3]_{Ur} = [NH_3]_{Bl} \qquad (3.32)$$

$$[NH_4^+]_{Ur} = \frac{[NH_3]_{Ur}}{[NH_3]_{Ur}/[NH_4^+]_{Ur}} \qquad (3.33)$$

Thus the difference in concentration of NH_4^+ across the blood urine barrier may be calculated.

Directions

1. Build and verify the model described in Figure 3.10.

Figure 3.10

Formulas

	A	B
1	Blood pH =	7.4
2	Urine pH =	6.5
3	Tot Blood Ammonia =	40
4		
5	BLOOD:	
6	[NH3]/[NH4] =	=10^(B1-9.25)
7	[NH3] =	=(B6/(B6+1))*B3
8	[NH4] =	=B3-B7
9	Tot Blood Ammonia =	=B7+B8
10		
11	URINE:	
12	[NH3]/[NH4] =	=10^(B2-9.25)
13	[NH3] =	=B7
14	[NH4] =	=B13/B12
15	Tot Urine Ammonia =	=B13+B14
16		
17	RATIO:	
18	Ur/Bl Ammonia =	=B15/B9

Values

	A	B
1	Blood pH =	7.4
2	Urine pH =	6.5
3	Tot Blood Ammonia =	40
4		
5	BLOOD:	
6	[NH3]/[NH4] =	0.0141
7	[NH3] =	0.5571
8	[NH4] =	39.4429
9	Tot Blood Ammonia =	40.0000
10		
11	URINE:	
12	[NH3]/[NH4] =	0.0018
13	[NH3] =	0.5571
14	[NH4] =	313.3057
15	Tot Urine Ammonia =	313.8629
16		
17	RATIO:	
18	Ur/Bl Ammonia =	7.8466

Chapter 3   Experiment 3.6

2. Notice that cell B9 repeats the input value in cell B3. This is done only to maintain an aesthetic symmetry in the table of blood and urine values.

3. All concentrations are in µM.

Questions

1. Explore the effect of urine pH on the concentration gradient of total ammonia ($NH_3 + NH_4^+$) across the blood urine barrier by completing the table that follows.

Total Blood Ammonia = 40 µM

Blood pH	Urine pH	Total Blood Ammonia	Total Urine Ammonia	Ratio Urine/Blood Ammonia
7.4	7.4	40		
7.4	7.2	40		
7.4	6.8	40		
7.4	6.4	40		
7.4	6.0	40		
7.4	5.6	40		
7.4	5.2	40		
7.4	4.8	40		
7.4	4.4	40		
7.4	4.0	40		
7.4	3.6	40		
7.4	3.2	40		

2. What kind of urine/blood total ammonia ratio would you expect if urine became more basic than blood? Test your prediction.

Problems

1. Explain in words that a bright high school student would understand: How is it possible for NH_4^+ to exist at different concentrations on the two sides of the renal tubule membrane when NH_3 and NH_4^+ are in equilibrium with each other on both sides?

2. Build a model in which the conjugate acid, rather than the conjugate base, may equilibrate across a membrane.

Chapter 4

Enzyme Kinetics

The rates of first-order, homogeneous chemical reactions are typically proportional to the concentration of reactant even at very high concentrations. As the concentration of reactant increases, the reaction velocity increases linearly.

Enzyme-catalyzed chemical reactions, on the other hand, are usually limited by the number of available catalytic sites. As the concentration of substrate increases, the reaction velocity asymptotically approaches a maximum velocity that cannot be exceeded without increasing the enzyme concentration.

A simple one-substrate, one-product, enzyme-catalyzed reaction (Figure 4.1) is described reasonably well by the simplifying assumptions of the Michaelis treatment:

1. The substrate concentration remains constant throughout the course of the reaction either because substrate concentration is much greater than enzyme concentration or because the substrate is continuously replenished.

2. The rate constants k_3 and k_4 are small with respect to k_1, k_2, k_5, and k_6 (causing the conversion of bound substrate to bound product to be rate-limiting and the concentration of EP to be effectively zero under "initial" conditions in the absence of product).

Figure 4.1

$$E + S \underset{k_2}{\overset{k_1}{\rightleftharpoons}} ES \underset{k_4}{\overset{k_3}{\rightleftharpoons}} EP \underset{k_6}{\overset{k_5}{\rightleftharpoons}} E + P$$

Chapter 4

These assumptions appear to satisfactorily approximate the actual situation in most simple enzymic reactions.

Assumption 2 of the Michaelis treatment is represented by Equations 4.1 and 4.2.

$$v = k_3[ES] \qquad (4.1)$$

Under initial conditions (when [EP] ~ 0):

$$[E]_{tot} = [E] + [ES] \qquad (4.2)$$

where $[E]_{tot}$ = total enzyme concentration

K_s, an equilibrium constant, expresses the affinity of substrate for enzyme (Equation 4.3).

$$K_s = \frac{[E][S]}{[ES]} \qquad (4.3)$$

K_s is defined as as a dissociation constant and therefore has the dimensions of concentration.

Substituting Equation 4.2 into 4.3 yields:

$$\frac{[ES]}{[E]_{tot}} = \frac{[S]}{K_s + [S]} \qquad (4.4)$$

Multiplying the left-hand side of Equation 4.4 by k_3/k_3 puts the Michaelis equation in its familiar form.

$$\frac{v}{V_{max}} = \frac{[S]}{K_s + [S]} = \frac{[S]}{K_m + [S]} \qquad (4.5)$$

where V_{max} = maximum attainable reaction velocity for a given value of $[E]_{tot}$
K_m = Michaelis constant (K_m is numerically equal to K_s in Michaelis kinetics [see below])

Enzyme Kinetics

Your textbook may use a different derivation (from Briggs and Haldane) in which K_m gets replaced by $(k_2 + k_3)/k_1$. You will see in Experiment 4.2 why we prefer the treatment presented.

From Equation 4.5 it can be seen that when v/V_{max} equals 1/2, $K_s = [S]$. The value of [S] at which $v/V_{max} = 1/2$ is termed the Michaelis constant K_m. In simple Michaelis kinetics, K_s and K_m always have the same value. K_m, however, is experimentally derivable from kinetic data whereas K_s is a thermodynamic equilibrium constant.

When the assumptions of the Michaelis treatment are valid, the rate of a reaction as a function of substrate concentration is determined by only two parameters:

1. The maximal velocity, V_{max}
2. The Michaelis constant, K_m

Other forms of the Michaelis equation that are useful include:

$$\frac{[S]}{K_m} = \frac{[ES]}{[E]} \tag{4.6}$$

$$[ES] = \frac{[E][S]}{K_m} \tag{4.7}$$

$$\frac{[ES]}{[E]} = \frac{[ES]}{[E]_{tot} - [ES]} = \frac{v}{V_{max} - v} \tag{4.8}$$

Equation 4.6 demonstrates that division of substrate concentration by the dissociation constant K_m yields the ratio of bound to unbound enzyme. This same relationship expressed as in Equation 4.7 is used extensively in upcoming exercises.

Equation 4.8 yields a term $v/(V_{max} - v)$ that is a linear function of [S] (see Equation 4.6). Thus laboratory data that lead to a linear plot of $v/(V_{max} - v)$ vs. [S] fit the assumptions of the Michaelis treatment. Nonlinear plots identify enzymic reactions that do not fit the assumptions of the Michaelis treatment (e.g., cooperative reactions).

Experiment 4.1

Eyring Kinetics and the Boltzmann Distribution

You may recall from elementary chemistry that Arrhenius proposed an empirical relationship to predict the effect of a change of temperature on reaction rates and that an exponential term in his expression was later interpreted as an activation energy. Eyring formalized this treatment (Equation 4.9) by suggesting that activation energy E_a represents the energy required by a molecule to reach a transition state in which it is as likely to proceed to products as it is to return to reactants (Figure 4.2).

$$k = A e^{\left(-\frac{E_a}{RT}\right)} \tag{4.9}$$

In this treatment the reaction velocity constant k is directly proportional to the fraction of reactant molecules with sufficient energy to reach the transition state. That fraction is estimated by using the Boltzmann relationship, the exponential term in Equation 4.9. E_a is the activation energy; R the gas constant (the Boltzmann constant in molar terms); T the absolute temperature; and A a proportionality factor that depends on such effects as the requirement for precise orientation of molecules during reaction.

Figure 4.2

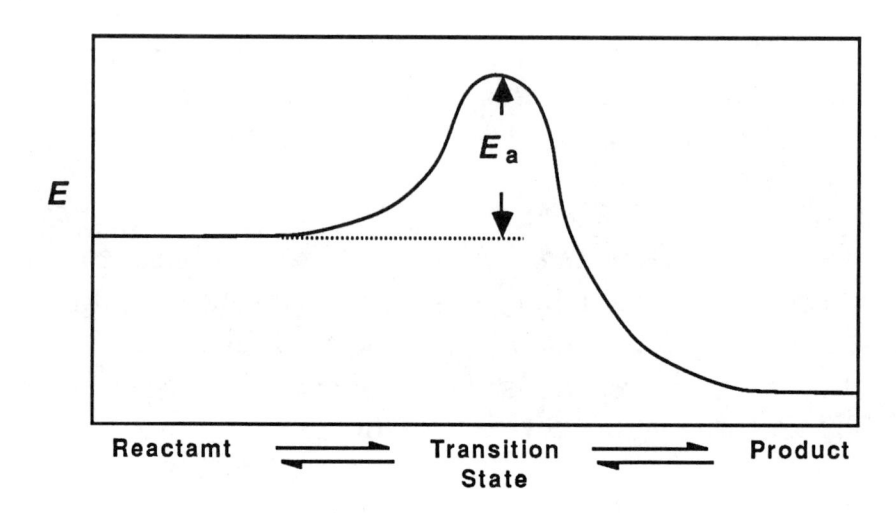

Before embarking on an exploration of enzyme kinetics, you should use this experiment to review these elementary considerations.

Directions

Build and verify the models described in Figure 4.3.

Figure 4.3

Formulas

	A	B	C	D	E
1	Variable is Ea			Variable is T	
2					
3	Ea1 =	15		Ea =	12
4	Ea2 =	5		R =	0.00199
5	R =	0.00199		T1 =	20
6	T =	25		T2 =	30
7					
8	k2/k1 =	=EXP((B3-B4)/(B5*(B6+273)))		k2/k1 =	=EXP((E3*(E6-E5))/(E4*(E5+273)*(E6+273)))

Values

	A	B	C	D	E
1	Variable is Ea			Variable is T	
2					
3	Ea1 =	15		Ea =	12
4	Ea2 =	5		R =	0.00199
5	R =	0.00199		T1 =	20
6	T =	25		T2 =	30
7					
8	k2/k1 =	21058981		k2/k1 =	1.9723

Chapter 4   Experiment 4.1

Questions

1. If you are not familiar with the Boltzmann relationship, use the left-hand side of the model described in Figure 4.3 to show the change in reaction rate that occurs with a change in E_a.

E_{a_1} (kcal/mol)	E_{a_2} (kcal/mol)	k_2/k_1
15.0	15.0	_____
15.0	14.5	_____
15.0	14.0	_____
15.0	13.5	_____
15.0	13.0	_____
15.0	12.5	_____

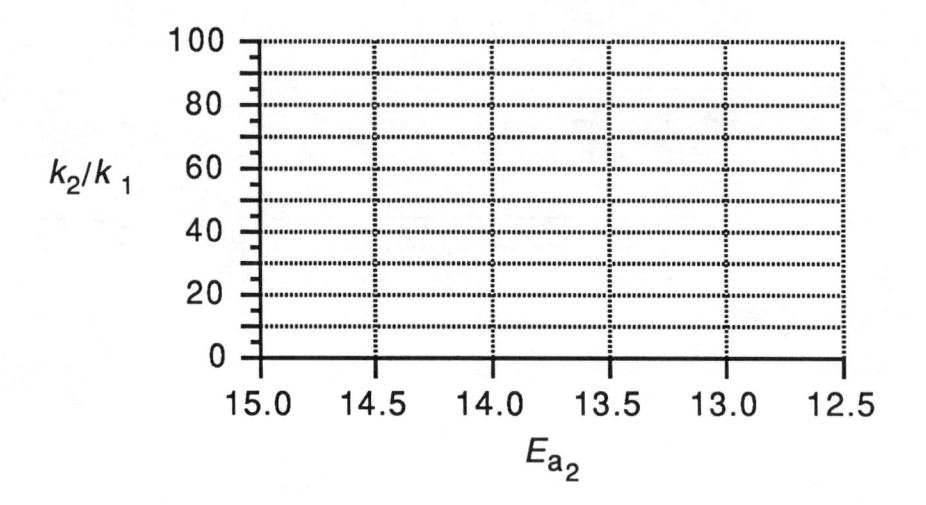

[Graph: k_2/k_1 vs E_{a_2}, y-axis 0 to 100, x-axis 15.0 to 12.5]

   a. What is the magnitude of change in E_a required to change the reaction rate 100-fold? 1000-fold? 4000-fold? 1,000,000-fold?

2. The change in reaction rate that results from an increase of 10°C in temperature is often called Q_{10}. Complete and graph the table below to explore the influence of E_a on Q_{10}.

Q_{10}

E_a (kcal/mole)	T = 10-20 °C	T = 40-50 °C	T = 80-90 °C
0.0	_____	_____	_____
0.2	_____	_____	_____
0.4	_____	_____	_____
0.6	_____	_____	_____
0.8	_____	_____	_____
1.0	_____	_____	_____

Q_{10}

E_a (kcal/mole)	T = 10-20 °C	T = 40-50 °C	T = 80-90 °C
1.2	————	————	————
1.4	————	————	————
1.6	————	————	————
1.8	————	————	————
2.0	————	————	————

△ 10 - 20 °C
▽ 40 - 50 °C
□ 80 - 90 °C

a. Would you expect Q_{10} values for enzymic reactions to be generally larger or smaller than for nonenzymic reactions? (**Hint:** How do enzymes influence E_a?)

3. Variation of A is not very interesting. The velocity constant is merely proportional to A (Equation 4.9).

Experiment 4.2

Assumptions of the Michaelis Treatment

You should begin your exploration of enzyme kinetic simulations by clarifying for yourself the implications of the assumptions that are the foundations of the Michaelis treatment. This is not normally done in biochemistry textbooks, but you have at your disposal a personal computer and an electronic spreadsheet program that makes such an exploration simple and straightforward.

Begin by reviewing the simplest model of a one-substrate, one-product, enzyme-catalyzed reaction below. In this model both substrate S and product P may associate with enzyme to form enzyme-ligand complex ES or EP. ES and EP are interconvertible via the making and breaking of covalent bonds.

$$E + S \underset{k_2}{\overset{k_1}{\rightleftharpoons}} ES \underset{k_4}{\overset{k_3}{\rightleftharpoons}} EP \underset{k_6}{\overset{k_5}{\rightleftharpoons}} E + P$$

In the original Michaelis treatment EP and its associated reactions were not considered (Figure 4.4a). The assumptions of the original Michaelis treatment were:

1. Constant [S] throughout the reaction.

2. Rapid association and dissociation of E and S to form ES and slow conversion of ES to E and P (i.e., $k_1, k_2 \gg k_3$). E, S, and ES are thus in virtual equilibrium.

If the original Michaelis treatment is modified to account for EP, then it also must be assumed that E, P, and EP are in virtual equilibrium (i.e., $k_1, k_2, k_5, k_6 \gg k_3, k_4$) (Figure 4.4b). This latter version of the Michaelis treatment was used to develop the mathematics at the beginning of this chapter.

Figure 4.4

$$\text{(a)} \quad E + S \underset{k_2}{\overset{k_1}{\rightleftharpoons}} ES \overset{k_3}{\longrightarrow} E + P$$

$$\text{(b)} \quad E + S \underset{k_2}{\overset{k_1}{\rightleftharpoons}} ES \underset{k_4}{\overset{k_3}{\rightleftharpoons}} EP \underset{k_6}{\overset{k_5}{\rightleftharpoons}} E + P$$

In the original Michaelis treatment EP was not considered (a). E and S were assumed to be in equilibrium with ES (k_1, $k_2 \gg k_3$ [displayed with heavy arrows above]).

If the original Michaelis treatment is modified to take EP into account E and S are assumed to be in equilibrium with ES (k_1, $k_2 \gg k_3$, k_4) and E and P are assumed to be in equilibrium with EP (k_5, $k_6 \gg k_3$, k_4 [displayed with heavy arrows above]).

In either case (Figure 4.4a, b) the resulting rate equations are identical (Equation 4.10) and K_m (the value of [S] at which $v = V_{max}/2$) equals K_s (the dissociation constant for E, S, and ES) equals k_2/k_1.

$$\frac{v}{v_{max}} = \frac{[S]}{K_m + [S]} \qquad (4.10)$$

An alternative approach, widely presented in biochemistry textbooks and called the Briggs-Haldane treatment, does not assume that enzyme, substrate, and enzyme-substrate complex are at equilibrium. Like the original Michaelis treatment, this approach ignores the existence of EP (Figure 4.5a).

If the original Briggs-Haldane treatment is modified to recognize the necessary existence of EP in the reaction pathway, it becomes obvious that to keep EP at insignificantly low concentrations it must be assumed that k_5 is much larger than k_6 (Figure 4.5b). This derivation leads to the conclusion that $K_m = (k_2 + k_3)/k_1$.

Figure 4.5

(a) $E + S \underset{k_2}{\overset{k_1}{\rightleftharpoons}} ES \overset{k_3}{\longrightarrow} E + P$

(b) $E + S \underset{k_2}{\overset{k_1}{\rightleftharpoons}} ES \underset{k_4}{\overset{k_3}{\rightleftharpoons}} EP \underset{k_6}{\overset{k_5}{\rightleftharpoons}} E + P$

In the Briggs-Haldane treatment EP was not considered (a). E and S were not assumed to be in equilibrium with ES (i.e., no assumptions with respect to the values of k_1, k_2, and k_3).

If the original Briggs-Haldane treatment is modified to take EP into account k_5 must be kept much larger than k_6 so that EP can never exist in appreciable concentrations (displayed with heavy arrow above).

In summary, the Michaelis treatment and the Briggs-Haldane treatment yield the same rate equation (Equation 4.10). However, the meaning attached to the parameter K_m differs in the two cases: In the Michaelis treatment, $K_m = k_2/k_1$ (the dissociation constant for the ES complex) and in the Briggs-Haldane treatment, $K_m = (k_2 + k_3)/k_1$.

Finally, if neither the Michaelis nor the Briggs-Haldane simplifications are assumed, then the rate equation for the enzymic reaction may be obtained from the expressions:

$$\frac{d[ES]}{dt} = k_1 [E][S] + k_4 [EP] - k_2 [ES] - k_3 [ES] \quad (4.11)$$

$$\frac{d[EP]}{dt} = k_3 [ES] + k_6 [E][P] - k_4 [EP] - k_5 [EP] \quad (4.12)$$

At steady state d[ES]/dt and d[EP]/dt are both equal to zero. After rather tedious algebra, a rate equation for the enzymic reaction can be derived that looks identical to Equation 4.10, except that K_m is replaced by K as defined in Equation 4.13.

$$K = [k_2(k_4 + k_5) + k_3 k_5]/[k_1(k_3 + k_4 + k_5)] \quad (4.13)$$

Assumptions of the Michaelis Treatment

In this experiment you will compare the outputs of Michaelis model and the Briggs-Haldane model with the general model when different assumptions are made about the rate constants of the individual steps.

Directions

Build and verify the model described in Figure 4.6.

Figure 4.6

Formulas

	A	B
1	k1 =	1
2	k2 =	2
3	k3 =	3
4	k4 =	4
5	k5 =	5
6	[S] =	6
7		
8	k2/k1 =	=B2/B1
9	(k2+k3)/k1 =	=(B2+B3)/B1
10	K =	=(B2*(B4+B5)+(B3*B5))/(B1*(B3+B4+B5))
11		
12	vel(k2/k1) =	=B6/(B6+B8)
13	vel((k2+k3)/k1) =	=B6/(B6+B9)
14	vel(K) =	=B6/(B6+B10)

Values

	A	B
1	k1 =	1
2	k2 =	2
3	k3 =	3
4	k4 =	4
5	k5 =	5
6	[S] =	6
7		
8	k2/k1 =	2.00
9	(k2+k3)/k1 =	5.00
10	K =	2.75
11		
12	vel(k2/k1) =	0.75
13	vel((k2+k3)/k1) =	0.55
14	vel(K) =	0.69

Chapter 4   Experiment 4.2

Questions

The rate equations for the general model, the Michaelis model, and the Briggs-Haldane model are identical but for the value computed for the Michaelis or Michaelis-like constant. Use the model described in Figure 4.6 to determine the conditions under which the Michaelis and Briggs-Haldane constants are equal or similar to the constant of the general model. Initial conditions (absence of product) are assumed, and so k_6 is not relevant.

Values of rate constants of 1.0 are used to designate fast reactions, and values of rate constants of 0.01 are used to designate slow reactions.

The table is designed such that conditions A, B, and C reflect assumptions in which one of the three steps (association/dissociation of substrate, reaction, and dissociation of product) is slow relative to the others. Conditions E, F, and G reflect assumptions in which one of the three steps is fast relative to the others.

Assumptions

	A	B	C	D	E	F	G
$k_1 =$	1.0	1.0	0.01	1.0	0.01	0.01	1.0
$k_2 =$	1.0	1.0	0.01	1.0	0.01	0.01	1.0
$k_3 =$	0.01	1.0	1.0	1.0	1.0	0.01	0.01
$k_4 =$	0.01	1.0	1.0	1.0	1.0	0.01	0.01
$k_5 =$	1.0	0.01	1.0	1.0	0.01	1.0	0.01
$k_2/k_1 =$	1	1	1	1	1	1	1
$(k_2+k_3)/k_1 =$	1.01	2	101	2	101	2	1
$K =$	1.00	0.15	34	1	34	1.97	0.67

Assumption A:

$$E + S \underset{k_2}{\overset{k_1}{\rightleftharpoons}} ES \underset{k_4}{\overset{k_3}{\rightleftharpoons}} EP \xrightarrow{k_5} E + P$$

By examining your completed table you should discover that this is the only assumption set in which all three methods of predicting reaction rate yield equivalent results. This one case is probably also the case that is at least approximately true for most enzyme-catalyzed reactions — association and dissociation of ligand is

Assumptions of the Michaelis Treatment

assumed to occur rapidly whereas the making and breaking of covalent bonds is assumed to be slower.

Assumption B:

$$E + S \underset{k_2}{\overset{k_1}{\rightleftharpoons}} ES \underset{k_4}{\overset{k_3}{\rightleftharpoons}} EP \overset{k_5}{\longrightarrow} E + P$$

Examination of your completed table should verify that neither the Briggs-Haldane or the Michaelis calculations yield rates equivalent to the general model.

Assumption C:

$$E + S \underset{k_2}{\overset{k_1}{\rightleftharpoons}} ES \underset{k_4}{\overset{k_3}{\rightleftharpoons}} EP \overset{k_5}{\longrightarrow} E + P$$

Does either the Michaelis or the Briggs-Haldane assumption sets yield results equivalent to the general model?

Assumption D:

$$E + S \underset{k_2}{\overset{k_1}{\rightleftharpoons}} ES \underset{k_4}{\overset{k_3}{\rightleftharpoons}} EP \overset{k_5}{\longrightarrow} E + P$$

The Michaelis-assumption set yields results equivalent to the general model when all rate constants have similar values. The Briggs-Haldane model, however, does not.

Assumption E:

$$E + S \underset{k_2}{\overset{k_1}{\rightleftharpoons}} ES \underset{k_4}{\overset{k_3}{\rightleftharpoons}} EP \overset{k_5}{\longrightarrow} E + P$$

Once again, the original Michaelis assumptions match the general model better than Briggs-Haldane.

Assumption F:

$$E + S \underset{k_2}{\overset{k_1}{\rightleftharpoons}} ES \underset{k_4}{\overset{k_3}{\rightleftharpoons}} EP \overset{k_5}{\longrightarrow} E + P$$

Chapter 4   Experiment 4.2

In this case Briggs-Haldane matches the general model with greater accuracy than the original Michaelis treatment.

Assumption G:

$$E + S \underset{k_2}{\overset{k_1}{\rightleftharpoons}} ES \underset{k_4}{\overset{k_3}{\rightleftharpoons}} EP \overset{k_5}{\rightarrow} E + P$$

Is this model realistic? Can either the Michaelis or the Briggs-Haldane treatments deal with it?

In summary, the original Michaelis treatment is satisfactory for the conditions assumed in column A of your table (conditions that are true fairly generally). The Briggs-Haldane treatment is also satisfactory in this case (because k_3 is negligibly small). The Briggs-Haldane model would be superior to the Michaelis treatment only under the conditions described in Column F — the assumption that interactions between enzyme and substrate are slow whereas those between enzyme and product are fast.

Many textbooks give the impression that the Briggs-Haldane constant is the "true" Michaelis constant and that the simplifying assumption that K_m equals the enzyme-substrate dissociation constant, K_s, can be used only in special cases. We hold the alternative view that the special case of column F — the only case in which the Briggs-Haldane treatment would be superior — probably occurs in nature very rarely if ever. You are now in a position to make up your own mind.

Problems

In your table, k_1 always equals k_2, and k_3 always equals k_4. You may wish to test conditions in which this is not true.

Conclusion

The general model described on page 72 and used throughout the discussion in this experiment is itself a simplified version of the catalytic activity of real enzymes. An enzyme probably goes through several complexes in the course of catalyzing a reaction, and thus the total enzyme concentration is partitioned between more than the two forms shown in the model. This simplification probably does not affect the qualitative conclusions of this experiment.

However, the statement

$$V_{max} = k_3[ES]_{max} = k_3[E]_{tot} \tag{4.14}$$

which is part of the classical Michaelis derivation, is not generally true. Since there must usually be other complexes after the initial ES complex, the enzyme at saturation is partitioned between these complexes and cannot be expected to exist solely as ES.

A better statement of Equation 4.14 would be

$$V_{max} = k_3[ES]_{max} \leq k_3[E]_{tot} \tag{4.15}$$

where in general we do not know how close $[ES]_{max}$ is to $[E]_{tot}$.

Experiment 4.3

Simple Michaelis Behavior

In addition to the assumptions and equations of the Michaelis treatment, you should understand the concept of momentary kinetic order before embarking on an exploration of simple Michaelis behavior.

Consider first the meaning of reaction order in homogeneous reactions (i.e., reactions that occur in a homogeneous environment and not at a surface). If the reaction rate increases linearly with the concentration (or more strictly, activity) of substrate S, the reaction is first-order with respect to S. If reaction rate increases linearly with $[S]^2$, the reaction is second-order with respect to S. If the reaction rate increases linearly with $[S]^n$, the reaction is nth-order with respect to S. The dependence of reaction rate on substrate concentration [S] and reaction order n may be written:

$$v = k[S]^n \tag{4.16}$$

where v = reaction rate
k = a proportionality constant

The slope m of v vs. [S] changes with [S] for all reaction orders other than 1 (Equation 4.17):

$$m = dv/d[S] = kn[S]^{n-1} \tag{4.17}$$

Thus a plot of v vs. [S] is linear only for reactions where $n = 1$ (Figure 4.7a). However, a plot of the logarithm of v vs. the logarithm of [S] is linear for uncomplicated, homogeneous reactions since Equation 4.18 is the equation of a straight line (Figure 4.7b). *The slope of this straight line is n, the reaction order.*

$$\begin{aligned} v &= k[S]n \\ \ln(v) &= \ln(k) + n\ln[S] \end{aligned} \tag{4.18}$$

Figure 4.7

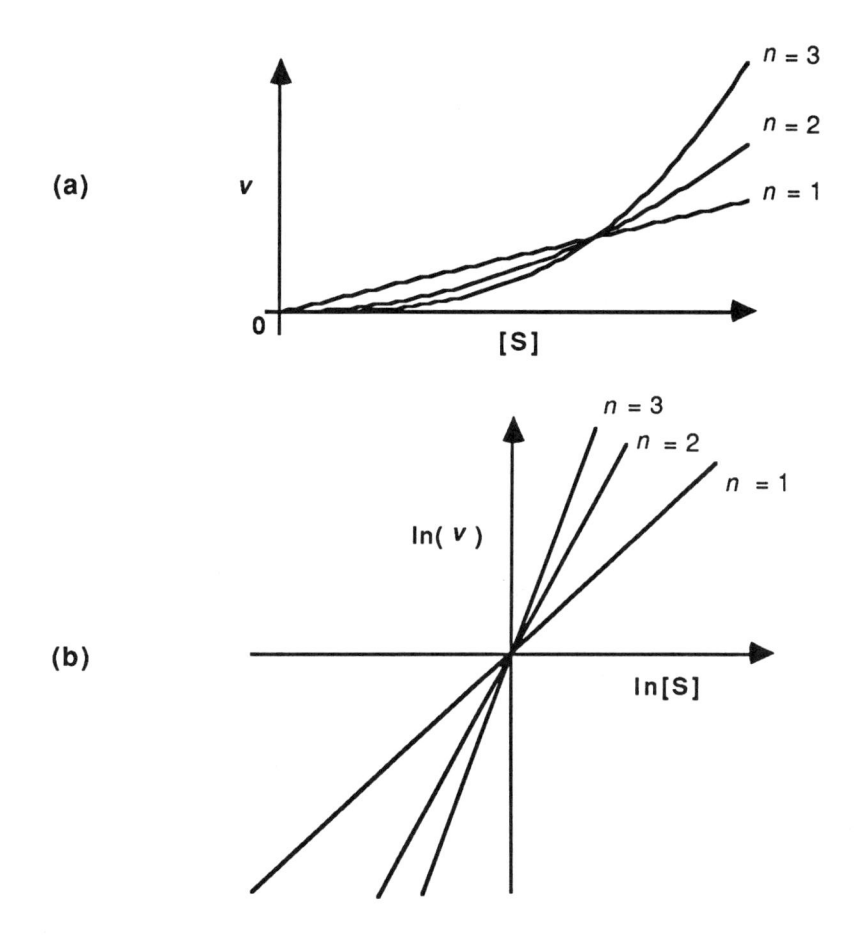

(a)

(b)

The reaction order *n* for homogeneous reactions is constant for all values of [S].

A plot of ln(*v*) vs. ln[S] is a straight line of slope *n* for homogeneous reactions (Figure 4.7b, Equation 4.18).

In graphic terms the meaning of reaction order in enzyme reactions is the same as in homogeneous reactions. Reaction order is the slope of the graph of the logarithm of v vs. the logarithm of [S]. However, because of the property of catalytic site saturation, the plot of ln(*v*) vs. ln[S] is not a straight line, and the slope varies as [S] changes. When simple Michaelis kinetics applies and when [S] is near 0, the slope of the plot of ln(*v*) vs. ln[S] is very nearly 1. As [S] approaches saturation values, the slope approaches 0 (Figure 4.8). Because reaction order varies with [S] and is not a constant property of the reaction (as it is in homogeneous reactions), reaction orders in enzymic reactions can be called *momentary kinetic orders*.

Chapter 4   Experiment 4.3

Figure 4.8

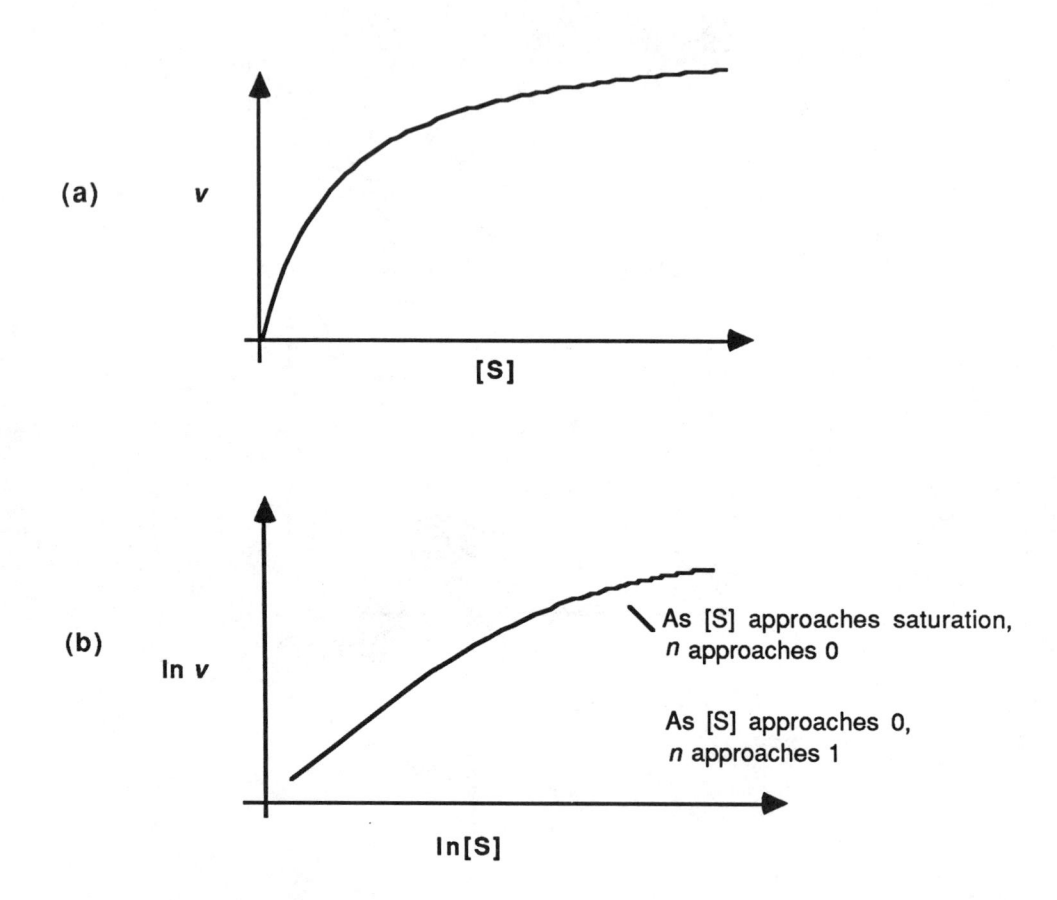

(a) v vs [S]

(b) ln v vs ln[S]

As [S] approaches saturation, n approaches 0

As [S] approaches 0, n approaches 1

Enzymic reactions have reaction rates that plateau as catalytic sites are saturated (a). Because of the phenomenon of site saturation plots of ln(v) vs. ln[S] are not linear; the slopes of such plots vary with [S] (b). Because the value of the slope of ln(v) vs. ln[S] may be interpreted as the reaction order, reaction order of enzymic reactions varies with [S] and can be called the momentary kinetic order.

Equations to determine the momentary kinetic order of an enzyme that exhibits Michaelis behavior can be derived algebraically (and is an exercise for you later in this chapter). The momentary kinetic order of any enzymic reaction can also be determined simply by approximating the tangent line of the plot of ln(v) vs. ln[S] by drawing a straight line through closely adjacent points (Figure 4.9).

Figure 4.9

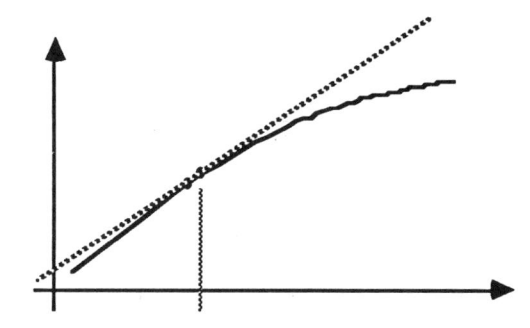

The slope of a straight line drawn between two closely spaced points on a curve approximates the slope of the curve in that region.

Directions

1. Build and verify the model described in Figure 4.10.

Figure 4.10

Formulas

	A	B
1	Km =	1
2	[S] =	2.5
3		
4	v/Vmax =	=B2/(B1+B2)
5	[S]/Km =	=B2/B1
6	v/(Vmax-v) =	=B4/(1-B4)
7	v'/Vmax =	=(1.001*B2)/(1.001*B2+B1)
8	Rxn Order =	=(LN(B7)-LN(B4))/(LN(1.001*B2)-LN(B2))

Values

	A	B
1	Km =	1
2	[S] =	2.5
3		
4	v/Vmax =	0.7143
5	[S]/Km =	2.5000
6	v/(Vmax-v) =	2.5000
7	v'/Vmax =	0.7145
8	Rxn Order =	0.2856

Chapter 4   Experiment 4.3

2. Cells B7 and B8 are used to calculate the momentary kinetic order.

   Cell B7 calculates v/V_{max} at a point close to that calculated in cell B4 (i.e., when [S] is incremented by 0.1%).

   Cell B8 calculates the slope (i.e., the momentary kinetic order) of the straight line drawn through two closely spaced points on the plot of $\ln(v/V_{max})$ vs. $\ln[S]$ defined by cells B4 and B7 from the equation:

$$\text{Reaction Order} = \frac{\ln(v'/V_{max}) - \ln(v/V_{max})}{\ln[S'] - \ln[S]} \quad (4.19)$$

   where [S'] = 1.001[S]
          v'/V_{max} = reaction rate at [S'] as a fraction of V_{max}

Questions

1. Complete and graph the table below:

$K_m = 1.0$

[S]	v/V_{max}	$v/(V_{max}-v)$	Reaction Order
0.00	0	0	
0.05	.05	.05	
0.10	.09	.1	.91
0.20	.17	.2	.83
0.50	.33	.5	.67
1.00	.5	1.0	.5
2.00	.67	2.0	.33
5.00	.83	5.0	.17
10.00	.91	10.0	.09

84

Simple Michaelis Behavior

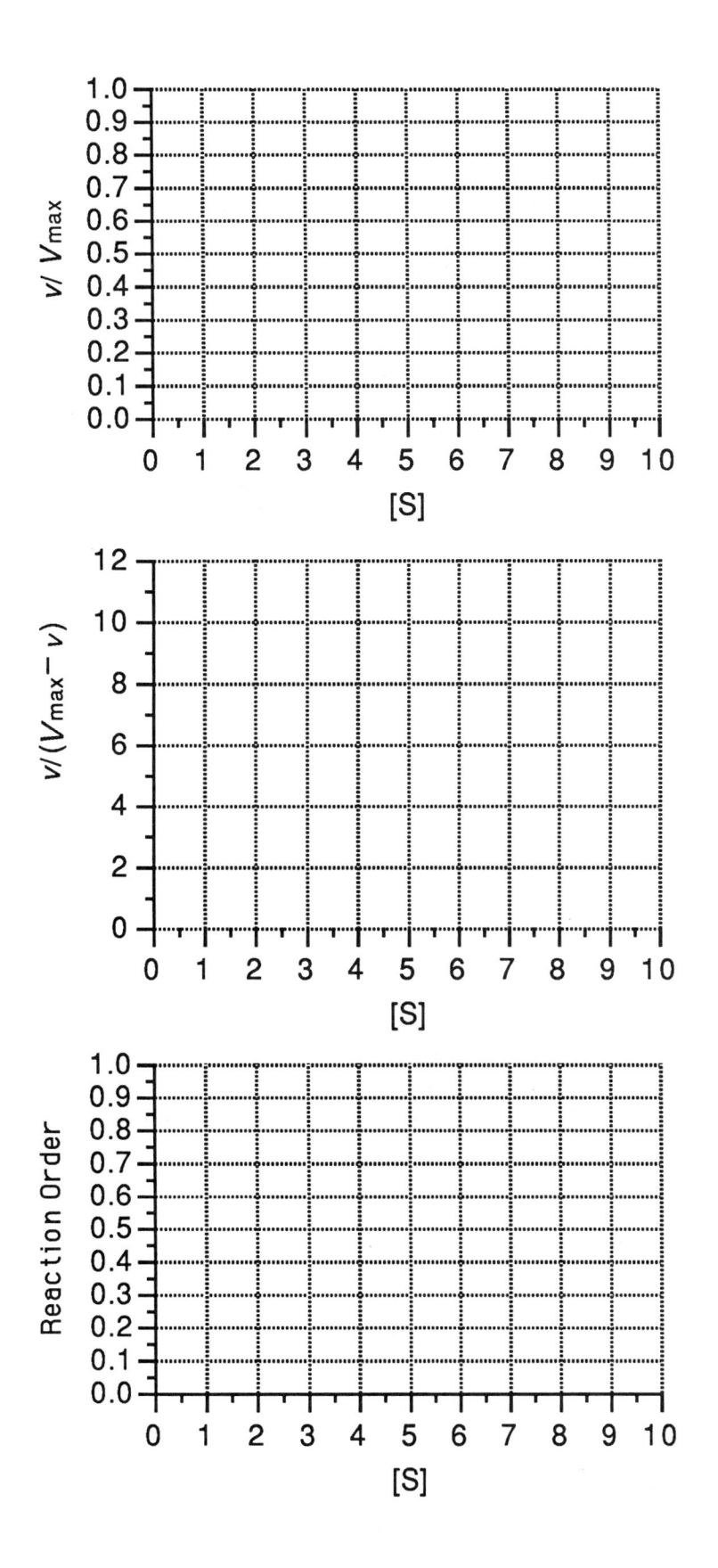

Chapter 4   Experiment 4.3

a. v/V_{max} vs. [S] yields the familiar Michaelis curve and shows how reaction rate plateaus as S saturates the catalytic site of the enzyme (i.e., [S] >> K_m).

b. For first-order enzyme reactions, $v/(V_{max} - v)$ vs. [S] (see Equation 4.8) is a linear function of [S]. In later experiments, dealing with cooperative enzyme reactions, plots of log [$v/(V_{max} - v)$] vs. log [S] are useful.

c. As [S] increases, what happens to the momentary kinetic order? Why?

d. The momentary kinetic order of a reaction of this type varies from 1 (at very low concentrations) to 0 (when [S] >> K_m). The table on page 84 illustrates a simple and probably unexpected relationship between the momentary kinetic order and the value of v/V_{max}. What is this relationship? You may find it interesting to derive the relationship mathematically.

2. Complete the table below and verify that the reaction rate is fixed by the ratio [S]/K_m. Why must this be so (remember Equation 4.6)?

K_m	[S]	[S]/K_m	v/V_{max}
0.5	1.25	2.15	.71
1.0	2.50	.71	2.5
2.0	5.00	.71	2.5
0.5	2.50	5	.83
1.0	5.00	5	.83
2.0	10.00	5	.83

3. Complete and graph the table on page 87 to examine the effect of variation of K_m on reaction rate. Variations in the affinity of substrate for enzyme (changes in the "effective" Michaelis constant) are among the most important factors in reaction-rate regulation in living cells.

Simple Michaelis Behavior

K_m	[S] = 2.5 v/V_{max}
0.00	_____
0.05	_____
0.10	_____
0.20	_____
0.50	_____
1.00	_____
2.00	_____
5.00	_____
10.00	_____

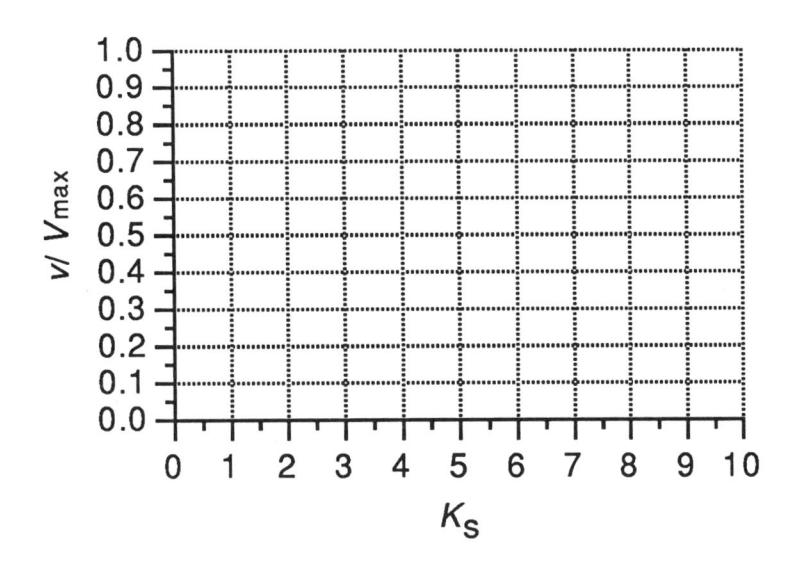

Experiment 4.4

Simple Michaelis Behavior: Catalytic Strength of an Enzyme

The initial reaction rates of enzymic reactions that satisfy the assumptions of the Michaelis treatment vary directly with k_3 and [ES] (Equation 4.1). When [S] is relatively low, [ES] varies inversely with K_m (Equation 4.7). As a consequence, the ratio k_3/K_m is sometimes used as a measure of the catalytic strength of an enzyme. It is often assumed that under otherwise constant conditions high values of k_3/K_m imply high reaction rates. In this experiment you will discover the conditions under which this is really so.

Directions

1. Build and verify the model described in Figure 4.11.

2. Cell B7 calculates the fraction of catalytic sites that bind substrate, and cell B8 uses this information to calculate reaction velocity v (Equations 4.4 and 4.5).

Figure 4.11

Formulas

	A	B
1	k3 =	0.01
2	Km =	3
3	Vmax =	1
4	[S] =	1
5		
6	k3/Km =	=B1/B2
7	[ES]/[E]tot =	=B4/(B4+B2)
8	v =	=B1*B3*B7
9	v/(k3/Km) =	=B8/B6

Values

	A	B
1	k3 =	0.01
2	Km =	3
3	Vmax =	1
4	[S] =	1
5		
6	k3/Km =	0.0033
7	[ES]/[E]tot =	0.2500
8	v =	0.0025
9	v/(k3/Km) =	0.7500

Catalytic Strength of an Enzyme

Questions

1. Use the model described in Figure 4.11 to complete the table below. Assume that V_{max} equals 1.

k_3	K_m	[S]	k_3/K_m	v	$v/(k_3/K_m)$
0.01	0.01	0.01	1	.005	.005
0.01	0.03	0.01	.33	.0025	.0075
0.01	0.10	0.01	.1	.0009	.009
0.01	0.30	0.01	.033	.0003	.0096
0.01	1.00	0.01	.01	.00	.01
0.01	3.00	0.01	.0033	.0000	.010
0.01	10.00	0.01	.0010	.000	.010
0.01	30.00	0.01	.0003	.000	.010
0.01	100.00	0.01	.0001	.000	.010
0.01	0.01	0.10			
0.01	0.03	0.10			
0.01	0.10	0.10			
0.01	0.30	0.10			
0.01	1.00	0.10			
0.01	3.00	0.10			
0.01	10.00	0.10			
0.01	30.00	0.10			
0.01	100.00	0.10			
0.01	0.01	1.00	.1	.0099	.0099
0.01	0.03	1.00	.33	.0097	.029
0.01	0.10	1.00	.100	.0091	.09
0.01	0.30	1.00	.033	.0077	.23
0.01	1.00	1.00	.01	.005	.50
0.01	3.00	1.00	.0030	.0025	.75
0.01	10.00	1.00	.0010	.0009	.909
0.01	30.00	1.00	.0003	.0003	.967
0.01	100.00	1.00	.0001	.0001	.9901

a. $v/(k_3/K_m)$ is a proportionality constant relating v and k_3/K_m. For each value of [S] in the table above, identify the range of values of K_m for which $v/(k_3/K_m)$ is approximately constant.

b. As K_m increases with respect to [S], does k_3/K_m become a better or worse measure of the catalytic strength of an enzyme?

Chapter 4 Experiment 4.4

2. Use the model described in Figure 4.11 to complete the table below.

k_3	K_m	[S]	k_3/K_m	v	$v/(k_3/K_m)$
0.01	0.01	0.01			
0.03	0.01	0.01			
0.10	0.01	0.01			
0.30	0.01	0.01			
1.00	0.01	0.01			
0.01	1.00	0.01			
0.03	1.00	0.01			
0.10	1.00	0.01			
0.30	1.00	0.01			
1.00	1.00	0.01			
0.01	0.01	1.00			
0.03	0.01	1.00			
0.10	0.01	1.00			
0.30	0.01	1.00			
1.00	0.01	1.00			

   a. Under what conditions is k_3 a good predictor of reaction rate?

3. Questions 1 and 2 should have lead you to the conclusion that k_3/K_m loses its correlation with reaction rate as $[S]/K_m$ increases, but that k_3 correlates with reaction rate under all conditions. Explain why this must be so.

4. Continue to explore the model described in Figure 4.11 until you are able to confirm this statement:

"Because the rate of a reaction is linearly proportional to k_3 at all concentrations of substrate, but rate is a nonlinear function of K_m, the ratio k_3/K_m can have no *general* significance. It is useful as a rough indicator of relative enzyme activity when $[S]/K_m$ is small."

Experiment 4.5

Equilibrium: Competition at a Catalytic Site

Consider a simple, first-order, homogeneous reaction:

$$A \underset{k_2}{\overset{k_1}{\rightleftarrows}} B$$

The forward reaction rate $k_1[A]$ is unaffected by the concentration of B. The reverse reaction rate $k_2[B]$ is unaffected by the concentration of A. Both forward and reverse reactions proceed independently of each other at all times. Equilibrium is simply the condition where the ratio of [B] to [A] is such that the forward and reverse reaction rates are equal.

In enzyme-catalyzed reactions the situation is quite different. Consider the one-substrate, one-product, enzyme-catalyzed reaction depicted below:

$$E + A \overset{K_a}{\rightleftarrows} EA \underset{k_4}{\overset{k_3}{\rightleftarrows}} EB \overset{K_b}{\rightleftarrows} E + B$$

The Michaelis treatment assumes that the two equilibria above (defined by the **dissociation** constants K_a and K_b) are rapid when compared with the interconversion of EA and EB (defined by the rate constants k_3 and k_4). Thus substrate A is in equilibrium with the enzyme-substrate complex EA; and product B is in equilibrium with the enzyme-product complex EB.

If A is reacting in the initial absence of B, the concentration of EB is nearly zero, and $v_{max(fwd)}$, the maximum forward reaction rate, occurs when the catalytic sites of the enzyme are saturated (Equation 4.20) and [EA] may be assumed to equal $[E]_{tot}$ (with the reservations expressed by Equation 4.15).

$$v_{max(fwd)} = k_3 [EA] = k_3 [E]_{tot} \qquad (4.20)$$

Similarly, if B is reacting in the initial absence of A, the concentration of EA is nearly zero, and $v_{max(rvs)}$, the maximum reverse reaction rate, occurs when the catalytic sites of the enzyme are saturated and [EB] is nearly equal to $[E]_{tot}$ (Equation 4.21).

Chapter 4   Experiment 4.5

$$v_{max(rvs)} = k_4 [EB] = k_4 [E]_{tot} \qquad (4.21)$$

If both A and B are present, however, neither can saturate the enzyme. For any given concentration of A, the fraction of A bound to enzyme is reduced by increasing concentrations of B. For any given concentration of B, the fraction of B bound to enzyme is reduced by increasing concentrations of A. Each ligand inhibits the rate of reaction of the other.

This is a *kinetic* inhibition of the individual rates of reaction. As either a homogeneous or an enzyme-catalyzed reaction approaches equilibrium, the net rate of conversion decreases because of the increasing rate of the reverse reaction. The *kinetic* inhibition of enzymic reactions by their products is an additional and quite different effect.

These same relationships apply if one enzyme catalyzes the reaction of two structurally related substrates (Figure 4.12).

A and B can inhibit the reaction of A' and B' and vice versa.

If v_{max} for the conversion of A' to B' is very low or zero, A' is simply a competitive inhibitor of the conversion of A to B.

The mathematics of competition at a catalytic site is straightforward and involves, once again, the construction of a partition equation.

Figure 4.12

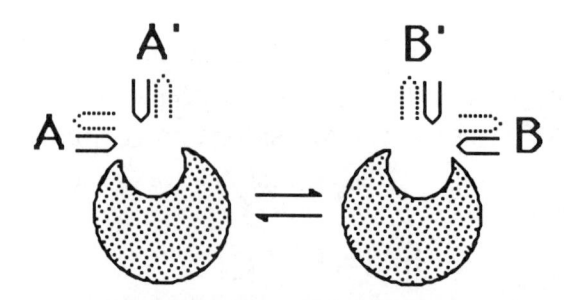

If an enzyme can catalyze the reaction of both A to B and A' to B', then A and B can inhibit the reactions of A' to B' and B' to A'. Similarly, A' and B' can inhibit the reactions of A to B and B to A.

Competition at a Catalytic Site

Consider the case described on page 91. Total enzyme is assumed to be the sum of free enzyme (E), enzyme-substrate complex (EA), and enzyme-product complex (EB) (Equation 4.22).

$$[E]_{tot} = [E] + [EA] + [EB] \tag{4.22}$$

Or, from Equation 4.7:

$$[E]_{tot} = [E](1 + [A]/K_a + [B]/K_b) \tag{4.23}$$

where K_a and K_b are the Michaelis constants for the substrates A and B, respectively.

Equation 4.23 may be rearranged to obtain the fraction of total enzyme that exists as EA and as EB (Equations 4.24 and 4.25).

$$\frac{[EA]}{[E]_{tot}} = \frac{[A]/K_a}{1 + \frac{[A]}{K_a} + \frac{[B]}{K_b}} \tag{4.24}$$

$$\frac{[EB]}{[E]_{tot}} = \frac{[B]/K_b}{1 + \frac{[A]}{K_a} + \frac{[B]}{K_b}} \tag{4.25}$$

Forward and reverse reaction rates may then be obtained from Equations 4.24, 4.25, and 4.1.

Equations that describe the behavior of a nonreacting competitive inhibitor may be obtained in a similar fashion.

Directions

Build and verify the model described in Figure 4.13.

Chapter 4   Experiment 4.5

Figure 4.13

Formulas

	A	B	C	D
1	[A] =	75	Ka =	1
2	[B] =	10	Kb =	0.1
3	[I] =	10	Ki =	0.1
4				
5			k3 =	5
6			k4 =	2
7				
8	Keq =	=(D2*D5)/(D1*D6)		% inhib.
9				
10	for [I] =	0	=B3	
11	[EA]/[E]tot =	Formula Set I	Formula Set II	
12	[EB]/[E]tot =			
13	[EC]/[E]tot =			
14	v(fwd) =	=D5*B11	=D5*C11	=100*(B14-C14)/B14
15	v(rvs) =	=D6*B12	=D6*C12	=100*(B15-C15)/B15
16	v(net) =	=B14-B15	=C14-C15	=100*(B16-C16)/B16

Values

	A	B	C	D
1	[A] =	75	Ka =	1
2	[B] =	10	Kb =	0.1
3	[I] =	10	Ki =	0.1
4				
5			k3 =	5
6			k4 =	2
7				
8	Keq =	0.25		% inhib.
9				
10	for [I] =	0	10	
11	[EA]/[E]tot =	0.4261	0.2717	
12	[EB]/[E]tot =	0.5682	0.3623	
13	[EC]/[E]tot =		0.3623	
14	v(fwd) =	2.1307	1.3587	36.2
15	v(rvs) =	1.1364	0.7246	36.2
16	v(net) =	0.9943	0.6341	36.2

Formula Set I
<u>Prototype Cell is B11</u>
=(B1/D1)/(1+(**B1/D1**)+(**B2/D2**))

Formula Set II
<u>Prototype Cell is C11</u>
=(B1/D1)/(1+(**B1/D1**)+(**B2/D2**)+(**B3/D3**))

Competition at a Catalytic Site

Questions

1. Use the model described in Figure 4.13 to complete and graph the table below (assume that no inhibitor is present).

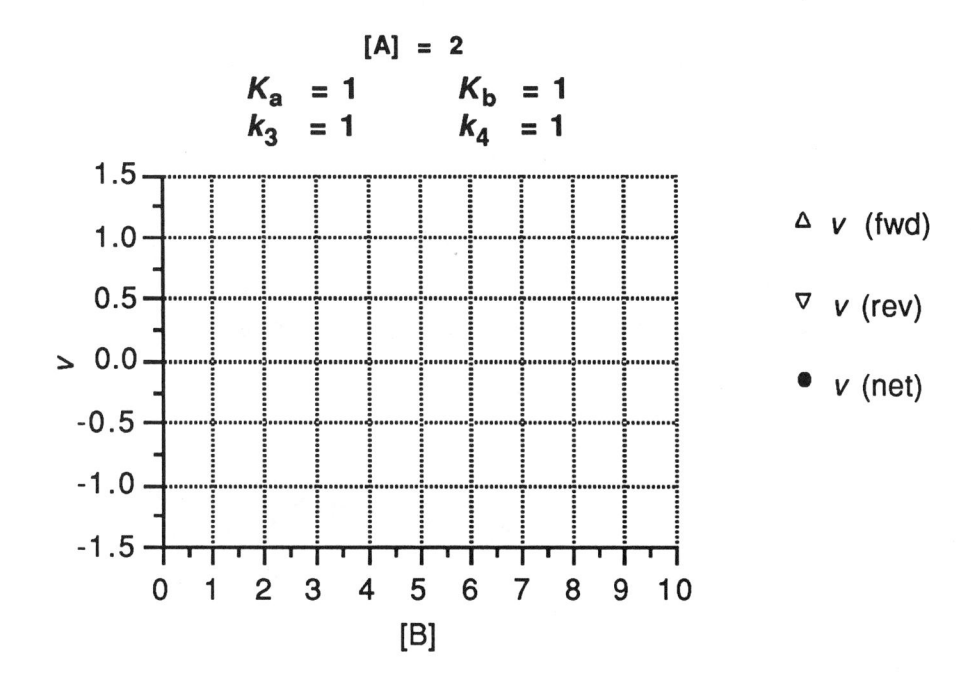

[A] = 2, $K_a = 1$, $K_b = 1$, $k_3 = 1$, $k_4 = 1$

[B]	v(fwd)	v(rvs)	v(net)
0.0	.66	0	.66
0.1	.64	.03	.61
0.2	.62	.06	.56
0.5	.57	.14	.42
1.0	.5	.25	.25
2.0	.4	.4	0
5.0	.25	.625	−.37
10.0	.15	.76	−.6

[A] = 2, $K_a = 5$, $K_b = 0.2$, $k_3 = 5$, $k_4 = 0.2$

[B]	v(fwd)	v(rvs)	v(net)
0.0	1.43	0	1.43
0.1	1.05	.05	1.0
0.2	.83	.08	.75
0.5	.51	.12	.38
1.0	.31	.15	.15
2.0	.175	.175	0
5.0	.07	.18	−.11
10.0	.04	.19	−.55

[A] = 2, $K_a = 1$, $K_b = 1$, $k_3 = 1$, $k_4 = 1$

△ v (fwd)
▽ v (rev)
● v (net)

95

Chapter 4   Experiment 4.5

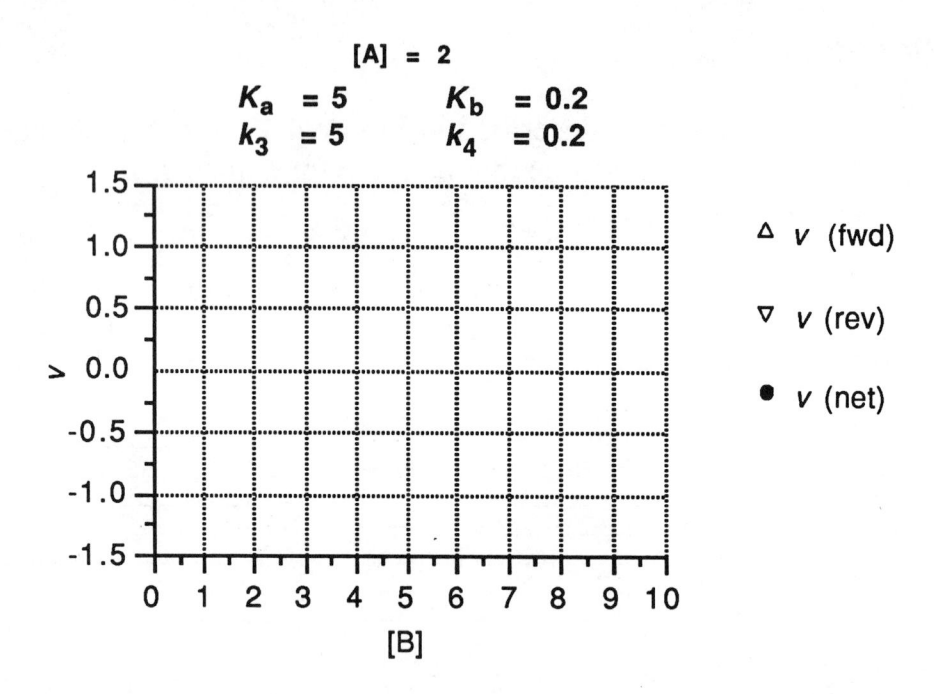

$[A] = 2$
$K_a = 5$ $K_b = 0.2$
$k_3 = 5$ $k_4 = 0.2$

△ v (fwd)

▽ v (rev)

● v (net)

(graph of v vs [B], v axis from -1.5 to 1.5, [B] axis from 0 to 10)

a. Examine the completed table on page 95. Under both specified conditions, what happens to the *forward* reaction rate v(fwd) as [B] increases? Would you expect this to occur if the reaction were homogeneous? Why or why not?

b. The forward reaction rates of enzymic reactions are inhibited by product because both substrate and product compete for the same catalytic site. Such inhibition is a *kinetic* inhibition of the reaction.

If K_a remains constant (or increases in value) while K_b decreases in value (i.e., if product competes more effectively for the catalytic site), predict the consequences on the ability of product to kinetically inhibit an enzymic reaction. Confirm your prediction by examining the table on page 95.

c. Explore other combinations of K_a and K_b to confirm your intuition with respect to the influence of the ratio of K_a to K_b on the kinetic inhibition of an enzyme reaction by product. Because the K_{eq} of a reaction is invariant at $k_3 K_b / k_4 K_a$ (i.e., fixed by the free energies of substrate and product) adjust the values of k_3 and k_4 so that K_{eq} does not change (see Equation 4.26).

Competition at a Catalytic Site

2. Use the model described in Figure 4.12 to complete and graph the table below. Note that since inhibitor is present, the reaction rates are displayed in column C of the model.

$$K_a = 1 \quad [A] = 1 \quad [B] = 1$$
$$K_b = 1 \quad K_i = 1$$
$$k_3 = 1 \quad k_4 = 1$$

[I]	v(fwd)	v(rvs)	v(net)	% Inhibition
0.0	.33	.33	0	0
0.1	.32	.32	0	3.2
0.3	.30	.30	0	9.1
1.0	.25	.25	0	25
3.0	.16	.16	0	50
10.0	.07	.07	0	76

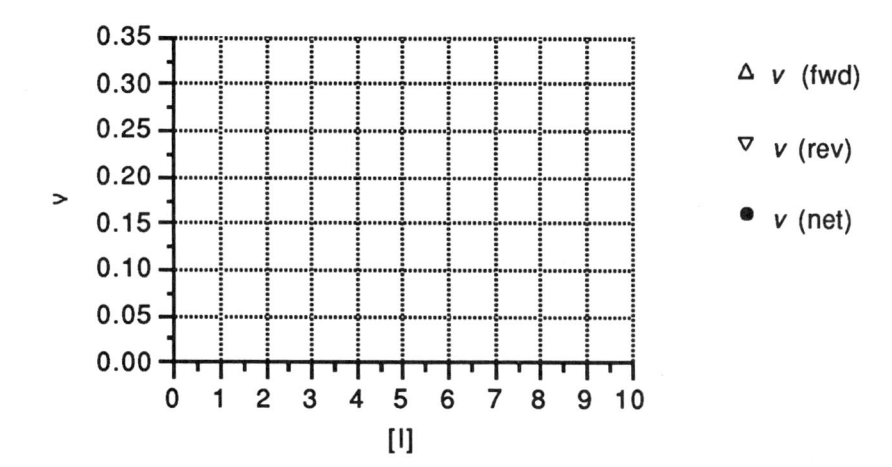

△ v (fwd)

▽ v (rev)

● v (net)

Beginning with the table and graph above, derive new tables and graphs by varying the rate and binding constants and answer the questions below.

a. What is the value of v(net) for a range of values of [I] when $K_a = K_b$, $k_3 = k_4$, and [A] = [B]?

b. What happens to the pattern of inhibition as the affinity of substrate for the catalytic site increases with respect to the affinity of product for the catalytic site (i.e., the K_s of substrate decreases with respect to the K_s of product) without change in the affinity of the inhibitor?

Chapter 4   Experiment 4.5

c. What happens to the pattern of inhibition as the affinity of inhibitor for the catalytic site increases with respect to the affinity of substrate and product for the catalytic site?

Work with the model described in Figure 4.13 until you are able to predict changes in inhibition patterns induced by changes in the rate and binding constants.

3. Complete and graph the table below and find the percent inhibition of $v(\text{fwd})$, $v(\text{rvs})$, and $v(\text{net})$ by using the model described in Figure 4.13.

[I] = 5

$K_a = 0.1$ $K_b = 10$ $K_i = 1.0$
k_3 and k_4 may assume any value

[I] = 5

$K_a = 0.1$ $K_b = 10$ $K_i = 0.1$
k_3 and k_4 may assume any value

[A]	[B]	v(fwd)	v(rvs)	v(net)	v(fwd)	v(rvs)	v(net)
10	0	4.7	—	4.7	33	—	33
8	2	5.8	5.8	5.8	38	38	38
6	4	7.5	7.5	7.5	44	44	44
5	5	8.8	8.8	8.8	49	49	49
4	6	10	10	10	54	54	54
2	8	18	18	18	69	69	69
0	10	—	71	71	—	96	96

$K_a = 0.1$ $K_b = 10$ $K_i = 1.0$
k_3 and k_4 at any value

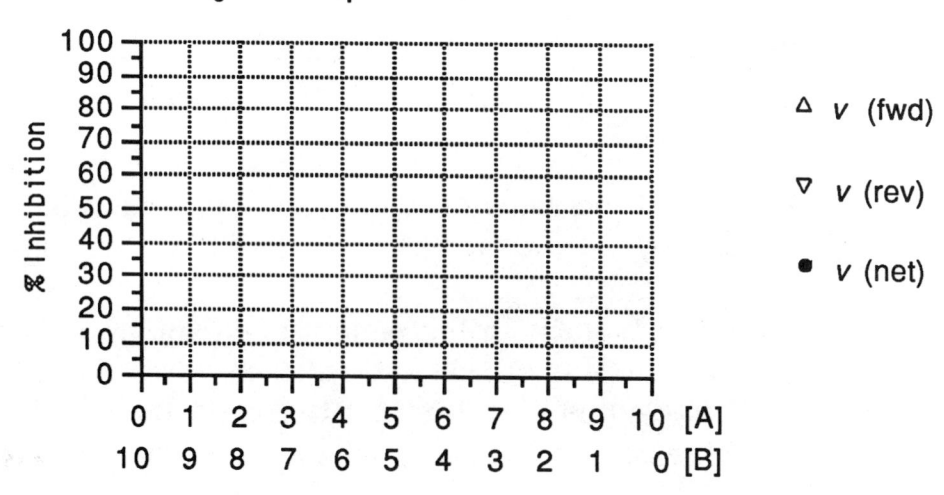

△ v (fwd)
▽ v (rev)
● v (net)

Competition at a Catalytic Site

$K_a = 0.1$ $K_b = 10$ $K_i = 0.1$
k_3 and k_4 at any value

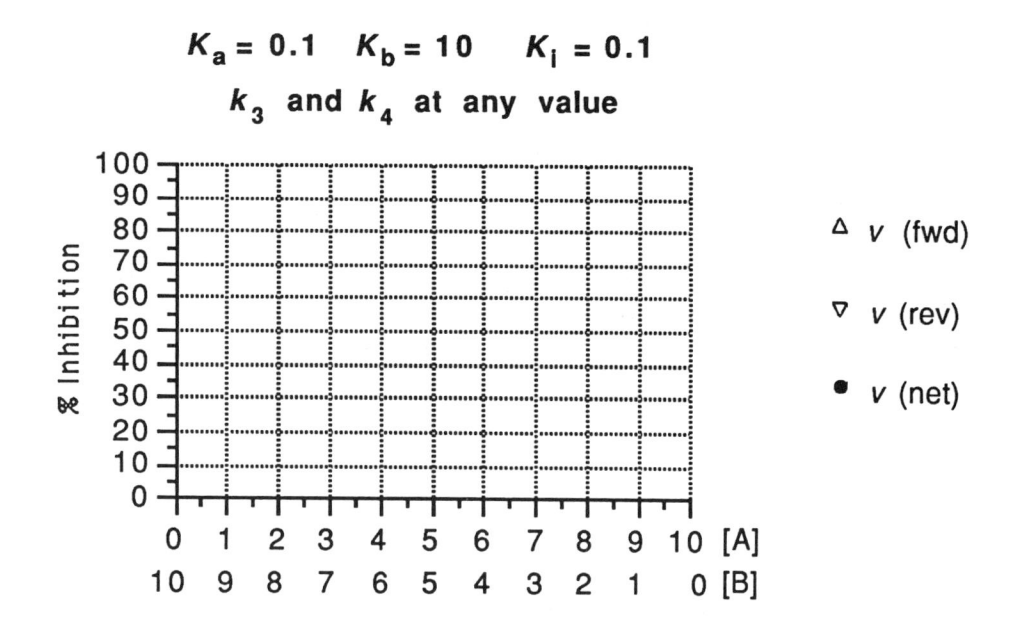

[Graph: % Inhibition vs [A] (0 to 10) and [B] (10 to 0), with legend:
△ v (fwd)
▽ v (rev)
● v (net)]

a. In these tables the extent of inhibition of the forward reaction in the absence of the product B is very different from the extent of inhibition of the reverse reaction in the absence of its product A. Why? Under what conditions would the two reactions be inhibited to the same extent?

b. Despite this difference, percent inhibition of v(fwd) is always equal to percent inhibition of v(rvs) for specific values of [A] and [B]. Why?

c. What happens to the pattern of inhibition if K_i changes from 1.0 to 0.1?

4. The model described in Figure 4.13 permits a wide range of explorations. Exploring various possibilities yields insights into aspects of the behavior of simple enzymes that are seldom considered in textbooks.

Problems

1. Derive the equations that exist in cells C11 through C16 of the model described in Figure 4.13.

Chapter 4   Experiment 4.5

2. The equilibrium constant for the conversion of A to B is:

$$K_{eq} = \frac{k_3 K_b}{k_4 K_a} \qquad (4.26)$$

K_{eq} can be altered by changing any of the rate or binding constants of the reaction. In the real world, however, K_{eq} is fixed by the nature of the reaction. Because changes in binding or in rate constants occur frequently by mutation, how do you explain the invariance of K_{eq} in nature?

3. Electronic spreadsheet programs often have the capability to create simple histograms. For example, many programs allow you to format a cell to display asterisks instead of numerical values (Figure 4.14).

Such capability, when it exists, adds an additional dimension to the capability of an electronic spreadsheet program to give you instant feedback on the status of a model.

If your program supports graphic capability, build and verify the model on page 101 and use it to explore this additional feature.

Figure 4.14

Formatted as a Number

	A	B
1	Value 1 =	3
2	Value 2 =	30
3	Value 3 =	15

Formatted as a Simple Histogram

	A	B
1	Value 1 =	***
2	Value 2 =	******************************
3	Value 3 =	***************

Competition at a Catalytic Site

The histogram values calculated in column D have been multiplied by the contents of a scaling factor in cell B4. The contents of cell B4 is arbitrary and is used only to generate a "convenient" number of asterisks. You may wish to use a different scaling factor to best fit the display of your computer and program.

Formulas

	A	B	C	D
1	[A] =	3	[E]/[E]tot =	=(1/(1+(B1/B6)+(B2/B7)+(B3/B8)))*B4
2	[B] =	2	[EA]/[E]tot =	=((B1/B6)/(1+(B1/B6)+(B2/B7)+(B3/B8)))*B4
3	[I] =	1	[EB]/[E]tot =	=((B2/B7)/(1+(B1/B6)+(B2/B7)+(B3/B8)))*B4
4	Scale =	15	[EI]/[E]tot =	=((B3/B8)/(1+(B1/B6)+(B2/B7)+(B3/B8)))*B4
5			v(fwd) =	=B9*D2
6	Ka =	0.1	v(rvs) =	=B10*D3
7	Kb =	0.2	v(net) =	=D5-D6
8	Ki =	0.3		
9	k3 =	5		
10	k4 =	2		
11	Keq =	=(B7*B9)/(B6*B10)		

Values

	A	B	C	D
1	[A] =	3	[E]/[E]tot =	0
2	[B] =	2	[EA]/[E]tot =	10
3	[I] =	1	[EB]/[E]tot =	3
4	Scale =	15	[EI]/[E]tot =	1
5			v(fwd) =	50.75
6	Ka =	0.1	v(rvs) =	6.77
7	Kb =	0.2	v(net) =	43.98
8	Ki =	0.3		
9	k3 =	5		
10	k4 =	2		
11	Keq =	5		

a. The affinity of the enzyme for the inhibitor plays an important role in determining the strength of inhibition. Vary k_1 in the model and observe this fact.

b. What is the significance of the ratio of $[I]/K_i$ to $[A]/K_a$ (or $[B]/K_b$)?

c. Vary other parameters in the model to improve your intuition. Remember that K_{eq} is fixed by the free energies of the reactant and product and (if you wish to observe allowable changes within the same reaction) must be kept constant by balancing changes in k_3, for example, with K_a.

Experiment 4.6

pH Response of an Enzyme

Many enzymic reactions are acid-base catalyzed. One functional group at the catalytic site donates a proton to the substrate, and another functional group accepts a proton from the substrate. In the simplest case, the reaction rate at a given substrate concentration should be proportional to the product of the concentration of the conjugate acid of the proton-donating group and the concentration of the conjugate base of the proton-accepting group (Figure 4.15).

Figure 4.15

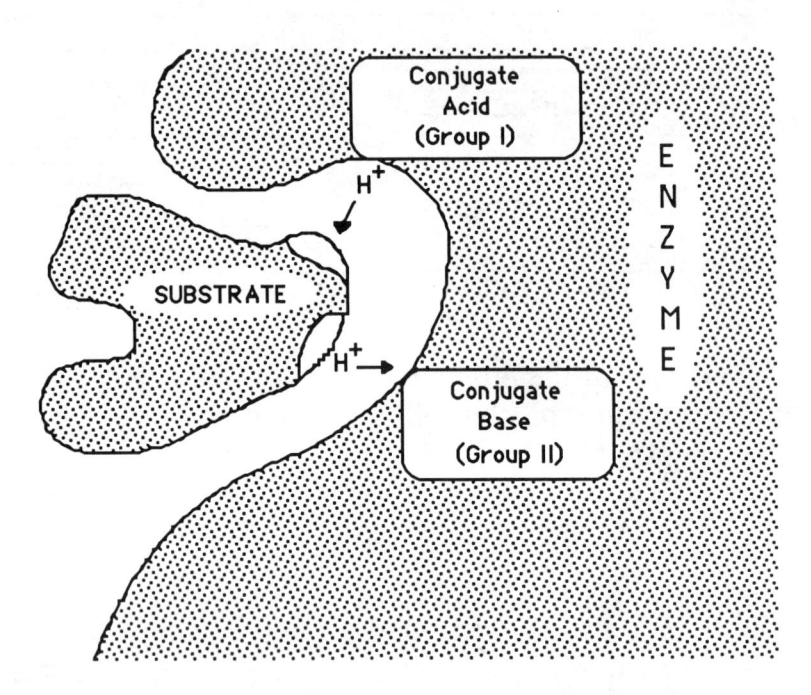

The substrate of an acid-base catalyzed enzymic reaction is converted to product when the conjugate acid of a proton-donating functional group on the enzyme (Group I) and the conjugate base of a proton-accepting functional group on the enzyme (Group II) are simultaneously presented to the substrate. In the simplest case, the probability of this occurrence may be assumed to be the product of the concentration of the conjugate acid of Group I times the concentration of the conjugate base of Group II. Thus the reaction rate is proportional to the product of these concentrations.

pH Response of an Enzyme

For each functional group the ratio of conjugate base to conjugate acid can be computed using the Henderson-Hasselbalch equation (Equations 4.27 through 4.28).

$$[B_1^-]/[HB_1] = 10^{(pH - pK_1)} \qquad (4.27)$$

$$[B_2^-]/[HB_2] = 10^{(pH - pK_2)} \qquad (4.28)$$

Where B_1^- = conjugate base of the proton-donating group
B_2^- = conjugate base of the proton-accepting group

Thus the mole fraction of the conjugate acid of the proton-donating functional group and the mole fraction of the conjugate base of the proton-accepting functional group may also be calculated (Equations 4.29 and 4.30).

$$\frac{[HB_1]}{[HB_1] + [B_1^-]} = \frac{1}{1 + [B_1^-]/[HB_1]} = \frac{1}{1 + 10^{(pH - pK_1)}} \qquad (4.29)$$

$$\frac{[B_2^-]}{[HB_2] + [B_2^-]} = \frac{1}{1 + [HB_2]/[B_2^-]} = \frac{1}{1 + 10^{(pK_2 - pH)}} \qquad (4.30)$$

Directions

1. Build and verify the models described in Figures 4.16 and 4.17.

Figure 4.16

Formulas

	A	B
1	pH =	7
2	pK(B1) =	7
3	pK(B2) =	5
4		
5	HB1/(HB1+B1) =	=1/((10^(B1-B2))+1)
6	B2/(HB2+B2) =	=1/((10^(B3-B1))+1)
7		
8	Rel. Activity =	=B5*B6

Values

	A	B
1	pH =	7
2	pK(B1) =	7
3	pK(B2) =	5
4		
5	HB1/(HB1+B1) =	0.5000
6	B2/(HB2+B2) =	0.9901
7		
8	Rel. Activity =	0.4950

Chapter 4   Experiment 4.6

Figure 4.17

Formulas

	A	B	C	D
1	pK(B1) =	7.5		
2	pK(B2) =	6.5		
3				
4	pH	HB1/(HB1+B1)	B2/HB2+B2	Rel. Activity
5	5.0	Formula Set I	Formula Set II	Formula Set III
6	5.5			
7	6.0			
16	10.5			
17	11.0			

Values

	A	B	C	D
1	pK(B1) =	7.5		
2	pK(B2) =	6.5		
3				
4	pH	HB1/(HB1+B1)	B2/HB2+B2	Rel. Activity
5	5.0	0.9968	0.0307	0.0306
6	5.5	0.9901	0.0909	0.0900
7	6.0	0.9693	0.2403	0.2329
16	10.5	0.0010	0.9999	0.0010
17	11.0	0.0003	1.0000	0.0003

Formula Set I
Prototype Cell is B5
=1/((10^(A5-**B1**))+1)

Formula Set II
Prototype Cell is C5
=1/((10^(**B2**-A5))+1)

Formula Set III
Prototype Cell is D5
=B5*C5

pH Response of an Enzyme

2. Figures 4.16 and 4.17 are essentially the same model. However, Figure 4.17 allows you to examine the behavior of a range of pHs simultaneously.

Questions

1. Use the model described in Figure 4.16 to complete and graph the table below:

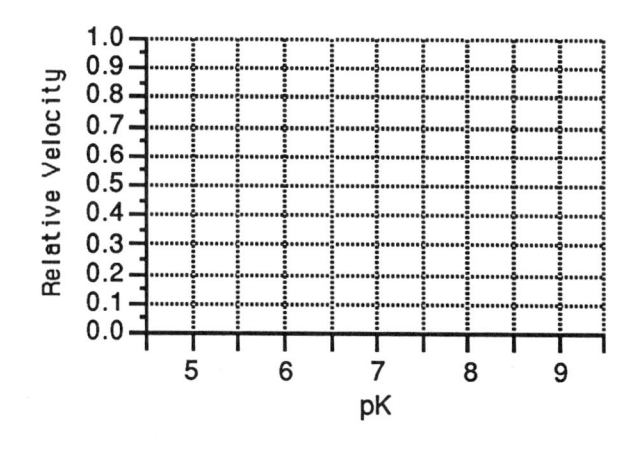

pH	pK(B_1)	pK(B_2)	Relative Velocity
7	7	5	.5
7	7	6	.45
7	7	7	.25
7	7	8	.05
7	7	9	.00
7	5	7	.00
7	6	7	.05
7	7	7	.25
7	8	7	.45
7	9	7	.5
7	8	6	.83
7	6	8	.01

△ pK of B_1 = 7 (B_2 is abscissa)

▽ pK of B_2 = 7 (B_1 is abscissa)

a. In your model, unlike real experiments, you may change the value of pK at will. At pH = 7, what happens to the relative velocity as the pK of the proton donor B_1 increases? As the pK of the proton acceptor B_2 increases?

b. The last two entries of the table above show that, contrary to what you might expect, enzyme activity is greater when the proton-donating functional group (B₁) is a weaker acid than the proton-accepting functional group (B₂). Why?

c. Explore the model described in Figure 4.16 until you can predict the direction of change of enzyme activity with changes in pH or the pKs of the functional groups.

2. The shape and width of the pH-response curve for an acid-base catalyzed enzymic reaction vary with the relative values of pK(B₁) and pK(B₂) and especially with the difference between them.

Use the model described in Figure 4.17 to investigate:

a. The pH response curve when pK(B₁) = pK(B₂) = 7.
When pK(B₁) = 8 and pK(B₂) = 6.
When pK(B₁) = 9 and pK(B₂) = 5.
When pK(B₁) = 10 and pK(B₂) = 4.
When pK(B₁) = 6 and pK(B₂) = 8.
When pK(B₁) = 5 and pK(B₂) = 9.
When pK(B₁) = 4 and pK(B₂) = 10.

b. What generalizations can you make about the results? How do the relative maximal reaction velocities vary with the pKs? How does the shape and width of the pH-response curve vary?

Problems

1. In many enzyme-catalyzed reactions, a specific ionic form of one or more substrates is required. Modify the model described in Figure 4.16 such that the substrate has an ionizable group — only one form of which is capable of entering into the reaction.

2. The pH-response curve of an enzyme-catalyzed reaction may result from effects other than those modeled in this experiment. Think about what these other effects might be and how they might interact with pK_1 and pK_2.

pH Response of an Enzyme

3. The model described in Figure 4.17 may be readily translated (in part) into a model that graphically represents the response of reaction rate to pH (Figure 4.18). Build and verify this model. *dyn 104*

This model has been normalized (cell B21) so that the longest bar will always be 50 units long. This allows you to compare the shapes of pH-response curves even when the peak velocities are very different. You may choose another value if you wish. If your program computes this model slowly, you may want to turn off automatic recalculation to facilitate entry of both variables prior to calculation.

Keep in mind as you work with this model that the longest bar may not necessarily represent peak response. Peak response might fall between the pH values in the table and be "in the cracks."

a. Compare the pH-response curve of an enzyme in which $pH(B_1) = 8.5$ and $pH(B_2) = 5.5$ with an enzyme in which $pH(B_1) = 5.5$ and $pH(B_2) = 8.5$. Does it change? What does change? (Consider the numerical answers in column B of Figure 4.20 as well as the histogram.)   *dyn 104 A*

If the shape of the pH-response curve suggests the participation of groups with pK values of 5.5 and 8.5, kinetic experiments will not determine which group is the proton donor and which is the proton acceptor. This ambiguity is a general feature of activity vs. pH curves.

b. Describe the pH response curves for the following values:

$pK(B_1)$	$pK(B_2)$	
8.0	6.0	*dyn 104 B*
7.5	6.5	
7.0	7.0	*dyn 104 C*
6.5	7.5	
6.0	8.0	*dyn 104 D*

It is intuitively obvious why the curve narrows as you progress through the first three pairs of values in the table. Why does the curve become broader again for the final two values?

Chapter 4   Experiment 4.6

Figure 4.18

Formulas

	A	B	C
1	8.5	= pK(B1)	
2	5.5	= pK(B2)	
3			
4	pH	Velocity	
5	3.5		=B5/B21
6	4.0		=B6/B21
7	4.5		=B7/B21
8	5.0	Formula Set I	=B8/B21
9	5.5		=B9/B21
10	6.0		=B10/B21
11	6.5		=B11/B21
12	7.0		=B12/B21
13	7.5		=B13/B21
14	8.0		=B14/B21
15	8.5		=B15/B21
16	9.0		=B16/B21
17	9.5		=B17/B21
18	10.0		=B18/B21
19	10.5		=B19/B21
20			
21		=MAX(B5..B19)/50	

Values

	A	B	C
1	8.5	= pK(B1)	
2	5.5	= pK(B2)	
3			
4	pH	Velocity	
5	3.5	0.0099	1
6	4.0	0.0307	2
7	4.5	0.0909	5
8	5.0	0.2402	13
9	5.5	0.4995	27
10	6.0	0.7574	40
11	6.5	0.9001	48
12	7.0	0.9396	50
13	7.5	0.9001	48
14	8.0	0.7574	40
15	8.5	0.4995	27
16	9.0	0.2402	13
17	9.5	0.0909	5
18	10.0	0.0307	2
19	10.5	0.0099	1
20			
21		0.0188	

Format Cells To Display Histogram

Formula Set I
Prototype Cell is B5
$=(1/(10^{\wedge}(A5-\mathbf{A1})+1))*(1/(10^{\wedge}(\mathbf{A2}-A5)+1))$

Experiment 4.7

Cooperative Kinetics: Changes in Substrate Affinity

Many enzymes, especially regulatory enzymes at metabolic branchpoints, catalyze reactions of kinetic order greater than one. Values between two and four are most common. Such enzymes have more than one catalytic site per enzyme molecule, and their kinetic behavior is usually caused by cooperative binding interactions (i.e., the binding of substrate at one catalytic site increases the binding affinity at other catalytic sites).

Consider the case in which cooperativity is essentially infinite. In this case, binding of substrate at one catalytic site changes the substrate affinity at all other catalytic sites such that they are immediately filled (Figure 4.19).

Figure 4.19

$$E + nS \xrightleftharpoons{K_s'} ES_n$$

When infinitely cooperative binding occurs at the catalytic sites of an enzyme molecule, the binding of one molecule of substrate causes all other sites to bind substrate very strongly and rapidly.

K_s' will be used to designate the equilibrium constant in this situation to distinguish it from K_s. K_s is the equilibrium constant that characterizes the binding of a single ligand to a single receptor site. As is conventional, K_s is a dissociation constant. Therefore K_s has the dimensions of concentration, and K_s' has the dimensions of concentration to the nth power, where n is the number of binding sites.

Under these conditions, only enzyme molecules with all binding sites empty and molecules with all binding sites filled will exist (Equation 4.31). Substituting Equation 4.31 into Equation 4.32 yields Equation 4.33.

$$[E]_{tot} = [E] + [ES_n] \tag{4.31}$$

$$K_s' = \frac{[E][S]^n}{[ES_n]} \tag{4.32}$$

$$\frac{v}{V_{max}} = \frac{[S]^n}{K_s' + [S]^n} \tag{4.33}$$

K_s' must have a different value than K_s, the equilibrium constant associated with the binding of a single molecule of substrate. Indeed, when cooperative binding prevails, the value of K_s varies with the number of substrate molecules already bound. Because K_s is not a constant under the terms of cooperative binding, it is appropriate to replace K_s by a symbol that does not imply constancy. When dealing with cooperative binding, K_s will always be replaced by the equivalent term $S_{0.5}$, representing the concentration of substrate at which reaction velocity equals $V_{max}/2$.

Thus Equation 4.33 is replaced by Equation 4.34:

$$\frac{v}{V_{max}} = \frac{[S]^n}{S_{0.5}^n + [S]^n} \tag{4.34}$$

Directions

1. Build and verify the model described in Figure 4.20.

2. For convenience in making comparisons the model computes the reaction rate of a cooperative enzymic reaction (expressed as v/V_{max}) as a function of substrate concentration at three combinations of values of $S_{0.5}$ and n.

3. You can save time during model building and value entry by turning off the automatic recalculation feature of your electronic spreadsheet program.

4. To decrease computation time, Equation 4.34 has been rewritten as:

$$\frac{v}{V_{max}} = \frac{1}{\left(\frac{S_{0.5}}{[S]}\right)^n + 1} \qquad (4.35)$$

Figure 4.20

Formulas

	A	B	C	D
1	n =	1	2	4
2	S(0.5) =	1	0.1	0.3
3				
4	[S]		v/Vmax	
5	1.E-8			
6	0.01			
7	0.02	Formula Set I	Formula Set II	Formula Set III
8	0.05			
9	0.1			
10	0.2			
11	0.5			
12	1			
13	2			
14	5			
15	10			
16	20			
17	50			

Values

	A	B	C	D
1	n =	1	2	4
2	S(0.5) =	1	0.1	0.3
3				
4	[S]		v/Vmax	
5	1.E-8	0.0000	0.0000	0.0000
6	0.01	0.0099	0.0099	0.0000
7	0.02	0.0196	0.0385	0.0000
8	0.05	0.0476	0.2000	0.0008
9	0.1	0.0909	0.5000	0.0122
10	0.2	0.1667	0.8000	0.1649
11	0.5	0.3333	0.9615	0.8853
12	1	0.5000	0.9901	0.9920
13	2	0.6667	0.9975	0.9995
14	5	0.8333	0.9996	1.0000
15	10	0.9091	0.9999	1.0000
16	20	0.9524	1.0000	1.0000
17	50	0.9804	1.0000	1.0000

Formula Set I
Prototype Cell is B5
=1/(1+((B2/A5)^B1))

Formula Set II
Prototype Cell is C5
=1/(1+((C2/A5)^C1))

Formula Set III
Prototype Cell is D5
=1/(1+((D2/A5)^D1))

Chapter 4   Experiment 4.7

Questions

1. Use the model described in Figure 4.20 to complete the table below and graph the region between [S] = 0 and [S] = 5.

v/V_{max}

[S]	$n=1$, $S_{0.5}=1$	$n=2$, $S_{0.5}=1$	$n=4$, $S_{0.5}=1$
10^{-8}	0	0	0
0.01	.01	0	0
0.02	.02	0	0
0.05	.048	0.002	0
0.10	.091	0.010	0.00
0.20	.167	0.038	0.002
0.50	.333	0.20	0.059
1	.500	0.50	0.50
2	.667	0.80	0.941
5	.833	0.962	0.998
10	.909	0.990	1.00
20	.952	0.998	1.00
50	.980	1.00	1.00

a. Explore your plot and note that, with infinite cooperativity assumed for the multisite enzymes, cooperative binding increases the sensitivity of reaction rate to changes in the concentration of substrate and that the sensitivity is much greater for an enzyme that contains four interacting sites than for one with two sites.

Changes in Substrate Affinity

2. Enzymes that have interacting substrate-binding sites, and thus catalyze reactions cooperatively, usually also have regulatory sites. When a specific modifier binds to the regulatory sites, the affinities of the catalytic sites for substrate (and thus $S_{0.5}$) are changed. Probably the most valuable metabolic consequence of a cooperative kinetic response is that the kinetic effect of a given change in the value of $S_{0.5}$ is very much greater if the enzyme is cooperative than if it is not.

   a. Complete and graph the table below to simulate a regulatory enzyme with a single binding site (and, of course, no cooperativity). The value of $S_{0.5}$ would normally be controlled by the concentration of a modifier molecule.

   At what concentrations of substrate is the effect of changed $S_{0.5}$ greatest?

   Can a modifier be an effective regulator of reaction rates if the concentration of substrate is high with respect to the possible range of values of $S_{0.5}$?

v/V_{max}

[S]	$n = 1$ $S_{0.5} = 0.1$	$n = 1$ $S_{0.5} = 0.2$	$n = 1$ $S_{0.5} = 0.4$
10^{-8}	0	0	0
0.01	.09	.04	.02
0.02	.17	.09	.05
0.05	.32	.20	.11
0.10	.50	.33	.20
0.20	.67	.50	.33
0.50	.83	.71	.55
1	.91	.83	.71
2	.95	.91	.83
5	.98	.96	.93

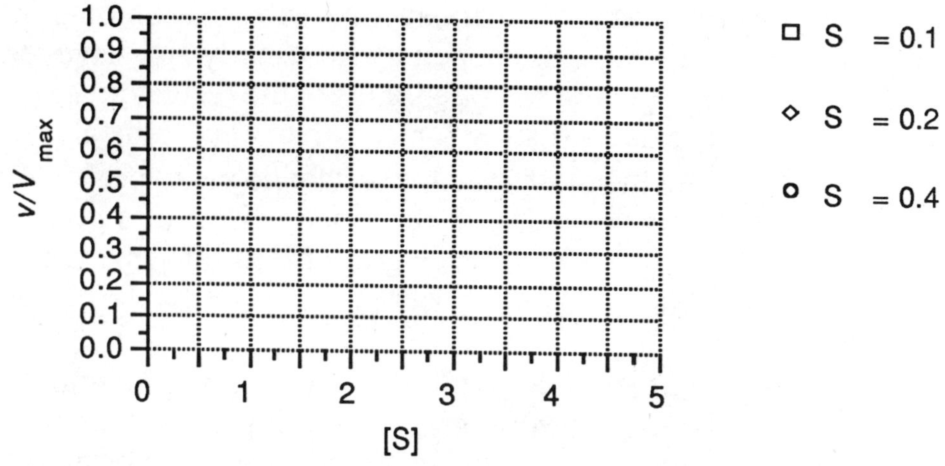

b. Now set $n = 2$ to simulate a cooperative, regulatory enzyme.

The change in velocity for a given change in substrate concentration is now much larger as $S_{0.5}$ varies over a fourfold range.

v/V_{max}

[S]	$n = 2$ $S_{0.5} = 0.1$	$n = 2$ $S_{0.5} = 0.2$	$n = 2$ $S_{0.5} = 0.4$
10^{-8}	______	______	______
0.01	______	______	______
0.02	______	______	______
0.05	______	______	______
0.10	______	______	______
0.20	______	______	______
0.50	______	______	______
1	______	______	______

dyn 114. pic

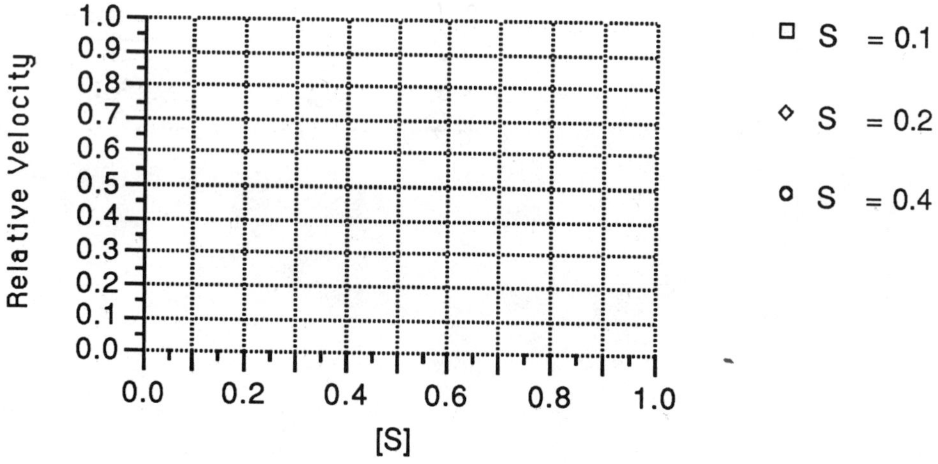

c. Set $n = 4$ to further increase the kinetic order (still assume infinite cooperativity). This simulation approximates the actual properties of many regulatory enzymes.

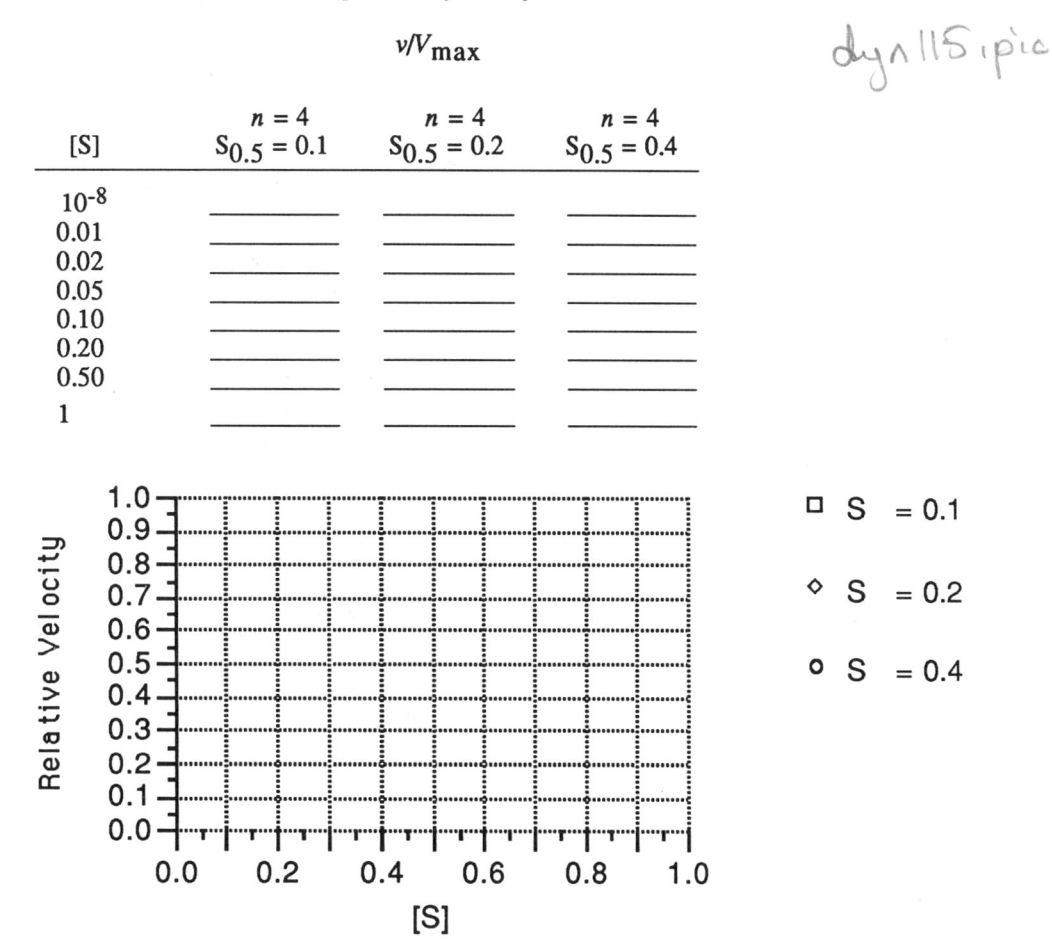

v/V_{max}

[S]	$n = 4$ $S_{0.5} = 0.1$	$n = 4$ $S_{0.5} = 0.2$	$n = 4$ $S_{0.5} = 0.4$
10^{-8}	_____	_____	_____
0.01	_____	_____	_____
0.02	_____	_____	_____
0.05	_____	_____	_____
0.10	_____	_____	_____
0.20	_____	_____	_____
0.50	_____	_____	_____
1	_____	_____	_____

□ S = 0.1
◇ S = 0.2
○ S = 0.4

d. Compare the results in (c) where $n = 4$ with the results in (a) where $n = 1$ and notice the enormous increase in sensitivity of reaction rate to changes in the value of $S_{0.5}$ resulting from cooperative binding of substrate. When [S] = 0.1, a fourfold change in affinity for substrate yields a change in reaction velocity of greater than 100-fold. When [S] = 0.2, a fourfold change in affinity causes the velocity to vary nearly across the whole range from zero to V_{max} (actually from 6% to 94% of V_{max}). This is probably the most valuable metabolic consequence of cooperativity.

e. Compare the results with those observed in (b) where $n = 2$.

Chapter 4   Experiment 4.7

You should now understand why most regulatory enzymes have evolved high-order cooperativity.

Problems

The model described in Figure 4.20 permits a wide range of explorations. More closely spaced values of n, $S_{0.5}$, or [S] can be used to investigate the properties of cooperativity at a higher degree of resolution. Increase the number of columns or change the spacing of concentration and explore. (If your electronic spreadsheet program is unacceptably slow in calculating this model, reduction of the number of rows will increase recalculation speed).

Experiment 4.8

Effective K_s

Although some modifiers change the maximum velocities of their target enzymes, most modifiers change the substrate affinity of the enzyme's catalytic sites. In this experiment an enzyme with one catalytic site and one regulatory site will be used to explore the relationship that exists between modifier, enzyme, and effective K_s. In the case of regulatory enzymes exhibiting *cooperative* kinetics, the sensitivity of V_{max} to changes in K_s is even greater (Experiments 4.7 and 4.10).

Consider an instance in which 50 out of a total of 100 enzyme molecules bind modifier. It might seem reasonable to expect that the apparent K_s of the assembly of 100 enzyme molecules would be a simple average of the two K_s values and that in general the apparent K_s would vary linearly with the amount of modifier-bound enzyme.

You may wish to ponder this issue further before conducting the following experiment.

Directions

1. Build and verify the model described in Figure 4.21. The assignable parameters are: [S]; K_s, the dissociation constant for substrate and enzyme with empty regulatory sites; K_s', the dissociation constant for substrate and enzyme with modifier-bound regulatory sites; and *f*, the fraction of enzyme molecules with filled regulatory sites.

2. The model first calculates the contributions to reaction velocity by enzyme molecules with empty modifier sites and by molecules with filled sites, using the Michaelis equation. The total velocity is calculated by summing the two contributions. The apparent, or

Chapter 4   Experiment 4.8

effective, K_s is then computed from the total velocity by rearranging the Michaelis equation:

$$K_s = \frac{[S](V_{max} - v)}{v} \qquad (4.36)$$

This is the value of K_s that, if all molecules in the solution behaved similarly, would give the same rate of reaction as is actually catalyzed by the mixture of enzyme molecules with and without bound modifier.

Finally, the effective K_s is normalized to give the observed change as a fraction of the maximal change that can be caused by bound modifier.

Figure 4.21

Formulas

	A	B	C	D	E	F
1	(Vmax = 100)					
2						
3	[S] =	1				
4	Ks =	5				
5	Ks' =	0.05				
6	(Modifier)					(Nrmlzd)
7	Frct bnd	vel	vel'	tot vel	Ks eff	Ks eff
8	0					
9	0.1					
10	0.2	Formula I	Formula II	Formula III	Formula IV	Formula V
11	0.3					
12	0.4					
13	0.5					
14	0.6					
15	0.7					
16	0.8					
17	0.9					
18	1					

(Continued.)

Figure 4.21 (Contd.)

Values

	A	B	C	D	E	F
1	(Vmax = 100)					
2						
3	[S] =	1				
4	Ks =	5				
5	Ks' =	0.05				
6	(Modifier)					(Nrmlzd)
7	Frct bnd	vel	vel'	tot vel	Ks eff	Ks eff
8	0	16.6667	0.0000	16.6667	5.0000	0.0000
9	0.1	15.0000	9.5238	24.5238	3.0777	0.3883
10	0.2	13.3333	19.0476	32.3810	2.0882	0.5882
11	0.3	11.6667	28.5714	40.2381	1.4852	0.7101
12	0.4	10.0000	38.0952	48.0952	1.0792	0.7921
13	0.5	8.3333	47.6190	55.9524	0.7872	0.8511
14	0.6	6.6667	57.1429	63.8095	0.5672	0.8955
15	0.7	5.0000	66.6667	71.6667	0.3953	0.9302
16	0.8	3.3333	76.1905	79.5238	0.2575	0.9581
17	0.9	1.6667	85.7143	87.3810	0.1444	0.9809
18	1	0.0000	95.2381	95.2381	0.0500	1.0000

Formula Set I
Prototype Cell is B8
=((1-A8)*100***B3**)/(**B4**+**B3**)

Formula Set II
Prototype Cell is C8
=(A8*100***B3**)/(**B5**+**B3**)

Formula Set III
Prototype Cell is D8
=B8+C8

Formula Set IV
Prototype Cell is E8
=((100-D8)***B3**)/D8

Formula Set V
Prototype Cell is F8
=(E8-**B4**)/(**B5**-**B4**)

Chapter 4   Experiment 4.8

Questions

1. Use the model described in Figure 4.21 to complete the table below.

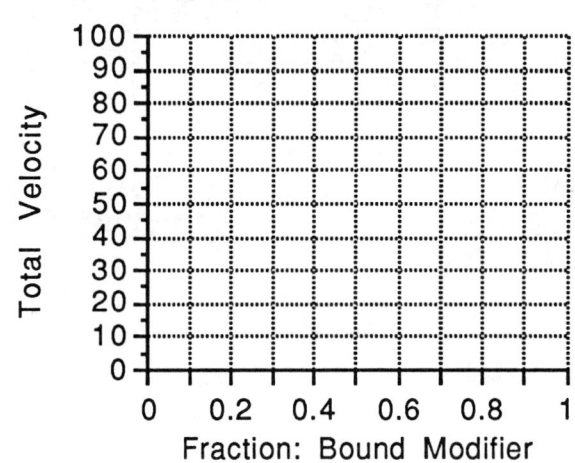

dyn 119 A.pic   *dyn 119.B.pic*   *dyn 119C.pic*

	[S] = 0.25 $K_S = 5$ $K_S' = 0.25$			[S] = 2.5 $K_S = 5$ $K_S' = 0.25$			[S] = 25 $K_S = 5$ $K_S' = 0.25$		
Fraction Bound	tot vel	K_s eff	(Normalized) K_s eff	tot vel	K_s eff	(Normalized) K_s eff	tot vel	K_s eff	(Normalized) K_s eff
0.0									
0.1									
0.2									
0.3									
0.4									
0.5									
0.6									
0.7									
0.8									
0.9									
1.0									

□ [S] = 0.25
◇ [S] = 2.5
○ [S] = 25

120

a. Examine your completed table and notice that the effective K_s is not proportional to the fraction of modifier-bound enzyme molecules. The effective value of K_s varies from K_s (when no modifier is bound) to K_s' (when all enzyme is modifier-bound). However, when [S] = 0.25, the effective K_s is midway between K_s and K_s' when only 10% of the enzyme molecules are bound to modifier. Explain.

What effect does [S] have on the relationship between modifier-bound enzyme and the values of effective K_s and total velocity? Explain.

b. Examination of the completed table shows that the fraction of modifier-bound enzyme is directly proportional to reaction velocity. How can this be when the modifier affects K_s and not V_{max}?

c. Must the incremental reaction velocity always be directionally proportional to the fraction of enzyme that bears modifier? Sketch a plot of reaction velocity against concentration of a positive modifier at constant substrate concentration. Can the dissociation constant of a modifier be validly estimated by use of such a plot?

2. Generate tables and plots similar to those of Question 1 for conditions in which $K_s = 0.25$ and $K_s' = 5$. Such conditions mimic the behavior of a negative modifier. Consider the similarities and differences in behavior of negative and positive modifiers.

Problems

Figure 4.21 permits you to generate automatically tables that display effective K_s as a function of the fraction of modifier-bound enzyme. Use this model to explore independently other aspects of the influence of K_s, K_s', and [S] on reaction velocity and effective K_s.

Conclusion

Nearly all real regulatory enzymes bind several molecules of substrate and several of modifier (usually cooperatively for both). The behavior of such enzymes is complex. This simple case, however, provides good preparation for consideration of the properties of more typical regulatory enzymes.

Experiment 4.9*

An Enzyme with Four Noninteracting Sites

Many enzymes have more than one catalytic site. In this experiment you will construct a model of an enzyme with four identical noninteracting catalytic sites. The behavior of such an enzyme is an excellent introduction to the complex, cooperative behavior of regulatory enzymes in which the sites do interact (the actual case for most regulatory enzymes). This subject will be covered in Experiment 4.10.

Because the binding sites of the model enzyme are identical, only five species of enzyme need be considered (Figure 4.22).

In this simplified treatment we will assume that the sum of these five possible species of enzyme is equal to the total enzyme concentration:

$$[E]_{tot} = [E] + [ES] + [ES_2] + [ES_3] + [ES_4] \qquad (4.37)$$

Remember, however, that other species of enzyme complexes that contain various combinations of product and substrate would have to be considered in a complete treatment (Experiment 4.5).

Figure 4.22

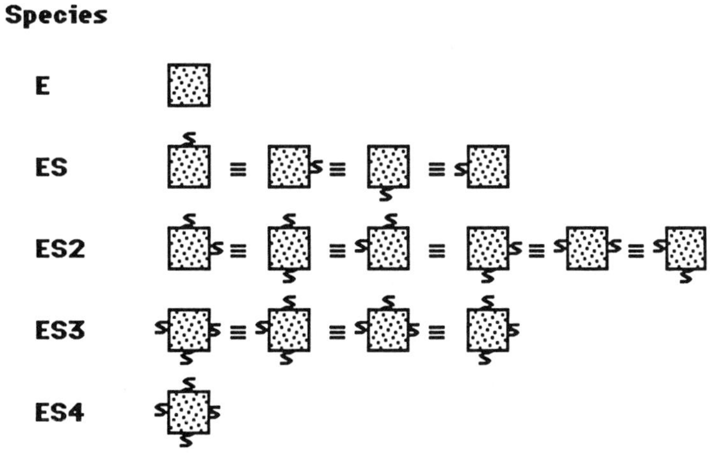

Equation 4.37 constitutes yet another example of a partition equation. Partition equations form the basis for most formal calculations of enzyme response vs. substrate (or other ligand) concentration. For cooperative enzymes, the equations become complex because of the increased number of terms that must be included in the equation and because of the calculations that determine the value of these terms.

In the present instance the value of each term may be calculated as follows:

$$[ES] = \frac{[E][S]}{K_1} \tag{4.38}$$

$$[ES_2] = \frac{[E][S]^2}{K_1 K_2} \tag{4.39}$$

$$[ES_3] = \frac{[E][S]^3}{K_1 K_2 K_3} \tag{4.40}$$

$$[ES_4] = \frac{[E][S]^4}{K_1 K_2 K_3 K_4} \tag{4.41}$$

K_1, K_2, K_3, and K_4 are the dissociation constants for ligand from ES, ES_2, ES_3, and ES_4, respectively.

Substituting Equations 4.38 through 4.41 into Equation 4.37 yields:

$$[E]_{tot} = [E]\left(1 + \frac{[S]}{K_1} + \frac{[S]^2}{K_1 K_2} + \frac{[S]^3}{K_1 K_2 K_3} + \frac{[S]^4}{K_1 K_2 K_3 K_4}\right) \tag{4.42}$$

or

$$\frac{[E]}{[E]_{tot}} = \frac{1}{\left(1 + \frac{[S]}{K_1} + \frac{[S]^2}{K_1 K_2} + \frac{[S]^3}{K_1 K_2 K_3} + \frac{[S]^4}{K_1 K_2 K_3 K_4}\right)} \tag{4.43}$$

Thus, given the substrate concentration and the appropriate equilibrium constants, $[E]/[E]_{tot}$ may be calculated from Equation 4.43. Once

An Enzyme with Four Noninteracting Sites

[E]/[E]$_{tot}$ is known, the relative concentrations of ES, ES$_2$, ES$_3$, and ES$_4$ may be calculated from Equations 4.38 through 4.41.

Because all sites are assumed to be identical, it might seem that K_1, K_2, K_3, and K_4 should have the same value and be equal to the intrinsic dissociation constant (the constant that would apply if each site were on a separate enzyme molecule). This is not true for statistical reasons.

Consider first K_{int}, the intrinsic dissociation constant (Figure 4.23).

K_{int} is equal to k_2/k_1, where k_1 and k_2 are the rate constants of association and dissociation. K_{int} is, by definition, the dissociation constant of the ligand-receptor complex when only one binding site exists.

For an enzyme with four free binding sites, the rate of binding to the enzyme is the sum of the rates at the four sites; thus it is four times as fast as binding to a single site. The rate of dissociation is the same as for an enzyme with only a single site. Thus K_1, the dissociation constant for the ES complex, is equal to $k_2/4k_1$ or $K_{int}/4$ (Figure 4.24).

Figure 4.23

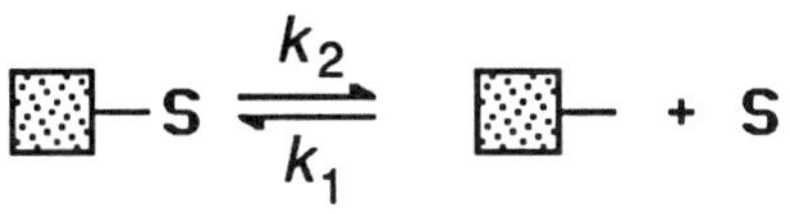

$$\boxed{}\!-\!S \underset{k_1}{\overset{k_2}{\rightleftharpoons}} \boxed{}\!-\! + S$$

Figure 4.24

$$\boxed{}\!-\!S \underset{4k_1}{\overset{k_2}{\rightleftharpoons}} \boxed{}\!-\! + S$$

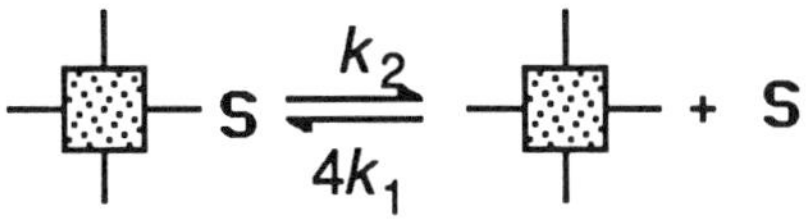

Such reasoning yields the values of all four equilibrium constants:

$$K_1 = K_{int}/4 \qquad (4.44)$$
$$K_2 = 2K_{int}/3 \qquad (4.45)$$
$$K_3 = 3K_{int}/2 \qquad (4.46)$$
$$K_4 = 4K_{int} \qquad (4.47)$$

where K_1, K_2, K_3, and K_4 are the dissociation constants for enzyme complexes containing one, two, three, and four molecules of substrate, respectively.

Directions

1. Build and verify the model described in Figure 4.25. The model computes [E], [ES], [ES$_2$], [ES$_3$], and [ES$_4$] as a fraction of total enzyme concentration and also calculates the fraction of sites that bind ligand at three different values of [S]. You may easily extend the model to compare more [S] values if you wish.

2. In Formula Set I, calculations such as [S]² and [S]³ have been written [S][S] and [S][S][S], respectively. In most electronic spreadsheet programs the latter calculations are quicker.

An Enzyme with Four Noninteracting Sites

Figure 4.25

Formulas

	A	B	C	D	E	F
1	Kint =	1.00	[S] =	0.5	1.0	2.0
2						
3	K1 =	=B1/4	[E] =	Formula Set I		
4	K2 =	=2*B1/3	[ES] =	Formula Set II		
5	K3 =	=3*B1/2	[ES2] =			
6	K4 =	=4*B1	[ES3] =			
7			[ES4] =			
8			Frtn Sites Bound =	Formula Set III		
9			Frtn Sites Bound =	Formula Set IV		
10			(Simple Michaelis)			

Values

	A	B	C	D	E	F
1	Kint =	1.00	[S] =	0.5	1.0	2.0
2						
3	K1 =	0.25	[E] =	0.1975	0.0625	0.0123
4	K2 =	0.67	[ES] =	0.3951	0.2500	0.0988
5	K3 =	1.50	[ES2] =	0.2963	0.3750	0.2963
6	K4 =	4.00	[ES3] =	0.0988	0.2500	0.3951
7			[ES4] =	0.0123	0.0625	0.1975
8			Frtn Sites Bound =	0.3333	0.5000	0.6667
9			Frtn Sites Bound =	0.3333	0.5000	0.6667
10			(Simple Michaelis)			

Formula Set I
Prototype Cell is D3
=1/(1+(D1/**B3**)+(D1*D1/(**B3*B4**))+(D1*D1*D1/(**B3*B4*B5**))
+(D1*D1*D1*D1/(**B3*B4*B5*B6**)))

Formula Set II
Prototype Cell is D4
=D3*(D**1**/**B3**)

Formula Set III
Prototype Cell is D8
=(D4+(2*D5)+(3*D6)+(4*D7))/4

Formula Set IV
Prototype Cell is D9
=(D1/(**B1**+D1))

Chapter 4   Experiment 4.9

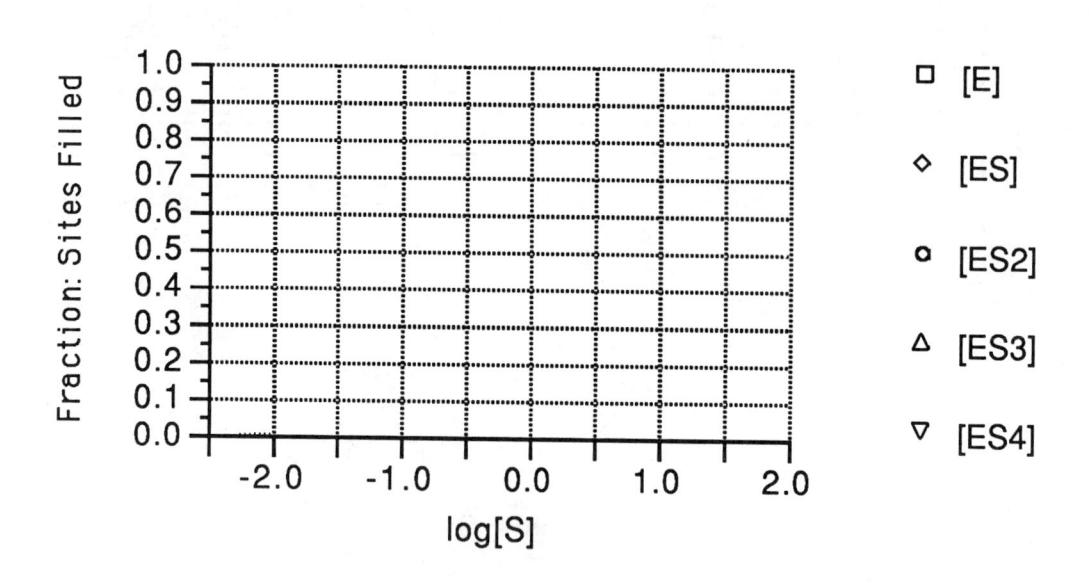

Questions

1. Use the model described in Figure 4.25 to complete and graph (as a semilog plot) the table below:

$K_{int} = 1$

Fraction of Sites Filled

[S]	log[S]	[E]	[ES]	[ES$_2$]	[ES$_3$]	[ES$_4$]	Total Sites	(Simple Michaelis) Total Sites
0								
0.01	-2.0							
0.0316	-1.5							
0.1	-1.0							
0.316	-0.5							
1	0.0							
3.16	0.5							
10	1.0							
31.6	1.5							
100	2.0							

□ [E]
◇ [ES]
○ [ES2]
△ [ES3]
▽ [ES4]

An Enzyme with Four Noninteracting Sites

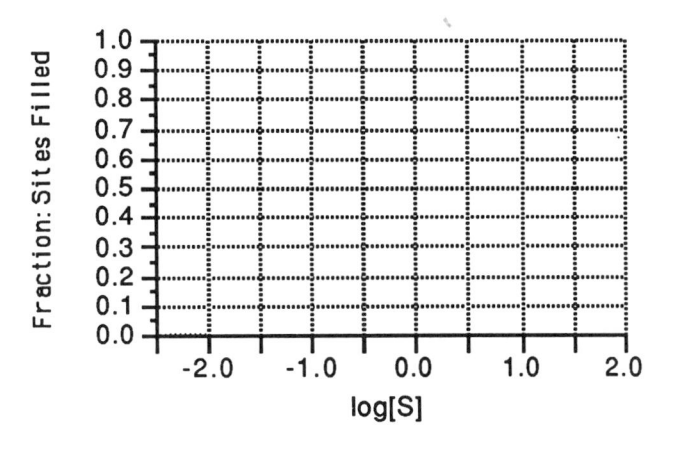

a. As [S] increases from 0 to 100, note how the concentration of each of the enzyme species E, ES, ES$_2$, ES$_3$, and ES$_4$ changes.

b. The total ligand bound at all four sites ([ES] + 2[ES$_2$] + 3[ES$_3$] + 4[ES$_4$]) increases smoothly with [S]. The curve is identical to the curve obtained for an enzyme with a single site (and also exhibits Michaelis kinetics).

2. For a simple enzyme with one catalytic site, the fraction of binding sites that bear ligand is a function of [S]/K_m (i.e., the fraction of enzyme bound to ligand does not change when [S] and K_m change proportionally). Complete the table below and determine whether the enzyme simulated in Figure 4.25 behaves similarly.

			Fraction Sites Filled	
[S]	K_m or K_{int}	[S]/K_m	Four-Site Enzyme	One-Site Enzyme
0.25	1	___	___	___
0.50	2	___	___	___
1.00	4	___	___	___
2.00	8	___	___	___
0.25	0.5	___	___	___
0.50	1	___	___	___
1.00	2	___	___	___
2.00	4	___	___	___

3. Your results should show that an enzyme with several noninteracting sites is kinetically indistinguishable from an enzyme with one site per molecule. Do you see why this must be so?

Although the statistical distributions that we deal with in this experiment have no kinetic consequences of interest in themselves,

Chapter 4   Experiment 4.9

they provide the basis for a similar treatment of cooperative enzymes. For cooperative enzymes such distribution patterns have important metabolic consequences.

Problems

1. Another way to explore the effect of [S] on an enzyme with four independent, catalytic sites is to prepare histograms of the relative fractions of [E], [ES], [ES$_2$], [ES$_3$], and [ES$_4$] at different values of [S]. Many electronic spreadsheet programs allow a special display format for cells that simulates a histogram. Other electronic spreadsheet programs have even more advanced, integrated graphics capabilities.

   If your electronic spreadsheet program supports such features, revise the model described in Figure 4.25 so that the values of [E], [ES], [ES$_2$], [ES$_3$], and [ES$_4$] are displayed as histograms (e.g., as shown below). Prepare histograms for [S] = 0.5, 1.0, and 2.0.

	A	B	C	D	E	F
1	Kint =	1.00	[S] =			1.0
2						
3	K1 =	0.25	[E] =	***		
4	K2 =	0.67	[ES] =	************		
5	K3 =	1.50	[ES2] =	******************		
6	K4 =	4.00	[ES3] =	************		
7			[ES4] =	***		
8	Scale =	50.00	Frtn Sites Bound =			50.0
9			Frtn Sites Bound =			50.0
10			(Simple Michaelis)			

2. When [S] = K_{int}, the concentrations of the five species of enzyme are in the ratio 1:4:6:4:1. These numbers are the coefficients of the terms resulting from the expansion of $(a + b)^4$ (consider Figure 4.22).

   If [S] ≠ K_{int} the concentrations of the five species of enzyme are the coefficients of the terms that result from the expansion of $(a + xb)^4$. What algebraic expression corresponds to x? Derive this expression by recalling the basic partition equation for a single site:

   $$[E]_{tot} = [E] + [ES] = [E](1 + [S]/K_m)$$

Experiment 4.10*

Cooperative Kinetics: Variable Degree of Cooperativity

Let's review the essential features of an enzyme with four independent catalytic sites (Experiment 4.9). If the dissociation rate constant k_1 equals the association rate constant k_2 at each of the four sites then, for statistical reasons, the effective dissociation constant depends on the number of ligand molecules currently bound (Figure 4.26).

In this experiment you will build a model of an enzyme with four catalytic sites that exhibit cooperative behavior (i.e., the affinity of a site for ligand is a function of the number of other sites at which ligand is bound).

To build this model, assume that the standard assumptions of the Michaelis treatment are valid and that the following additional assumptions also hold:

1. The probability that bound substrate will undergo catalysis is independent of whether other catalytic sites contain substrate. (This simply means that binding is cooperative, but the actual catalysis of the reaction is not.)

Figure 4.26

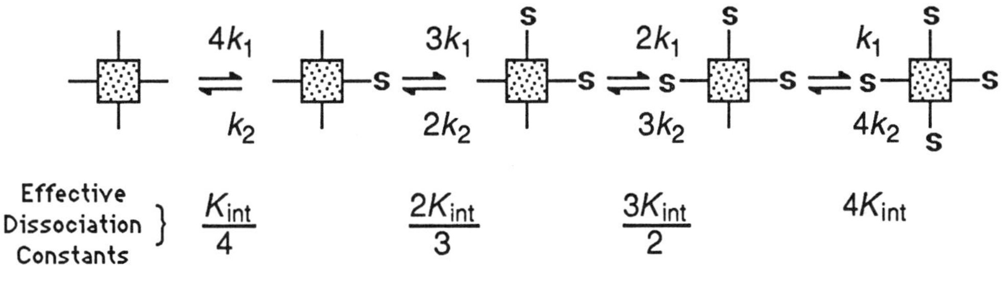

K_{int} is the intrinsic dissociation constant of each catalytic site (k_2/k_1)

Chapter 4   Experiment 4.10*

2. Each molecule of bound substrate affects the free energy of binding (i.e., the effective K_s [or K_m, when Michaelis kinetics is assumed]) at all other catalytic sites identically.

For example, if the binding of one mole of ligand decreases the free energy of binding (i.e., increases the negative value of ΔG_{bind} and hence the strength of binding) at other sites by x kcal and the effective K_s by a factor α, then the binding of two moles of ligand will increase the strength of binding by $2x$ kcal (and the effective K_s by a factor α^2) at each of the remaining two sites.

It is unlikely that these assumptions are precisely true, but their deviations from reality are probably not serious.

Before proceeding any further we must stop and reconsider our use of the notation K_s (or K_m). In this experiment and in the experiments that follow you will be exploring cooperative enzymes. The strength of binding of ligand to enzyme is variable. Thus K_s (with its implication of constant value) is a serious misnomer. *From this point on in our discussions of cooperative enzymes, K_m will be replaced by $S_{0.5}$.* $S_{0.5}$ is the concentration of substrate at which the reaction velocity equals $V_{max}/2$. It is independent of mechanism, and its use avoids the connotation of constant value. For a simple Michaelis enzyme, $S_{0.5}$ has the same value as K_s or K_m.

Under the circumstances defined prior to our slight digression, the effective dissociation constant of each enzyme form (K_1, K_2, K_3, and K_4) will be as displayed in Figure 4.27.

Figure 4.27

$$E \underset{\frac{K_{int}}{4}}{\overset{K_1}{\rightleftharpoons}} E\text{-}S \underset{\frac{2\alpha K_{int}}{3}}{\overset{K_2}{\rightleftharpoons}} S\text{-}E\text{-}S \underset{\frac{3\alpha^2 K_{int}}{2}}{\overset{K_3}{\rightleftharpoons}} S\text{-}E\text{-}S \underset{4\alpha^3 K_{int} S}{\overset{K_4}{\rightleftharpoons}} S\text{-}E\text{-}S$$

The overall equilibrium constant is the product of the individual equilibrium constants, $K = K_1 K_2 K_3 K_4$. For a simple case in which the value of α is constant for all sites, the value of $S_{0.5}$ is $K^{1/4}$.

Thus

$$S_{0.5} = (K_1 K_2 K_3 K_4)^{1/4}$$

$$= \left[\left(\frac{K_{int}}{4} \right) \left(\frac{2 \alpha K_{int}}{3} \right) \left(\frac{3 \alpha^2 K_{int}}{2} \right) 4 \alpha^3 K_{int} \right]^{1/4} \quad (4.48)$$

$$= K_{int} \, \alpha^{1.5}$$

or

$$K_{int} = \frac{S_{0.5}}{\alpha^{1.5}} \quad (4.49)$$

Two final points:

1. The change in the free energy of binding, ΔG_{bind}, that occurs because of the binding of ligand at one other site will be identified as $\Delta\Delta G_{bind}$.

2. α may be derived from $\Delta\Delta G_{bind}$ by use of the relationship $\Delta\Delta G_{bind} = RT \ln(\alpha)$. The absence of a negative sign on the right side of the equation is a consequence of our definition of α.

Directions

1. Build and verify the model described in Figure 4.28

2. Compare the model with that described for an enzyme with four independent catalytic sites (Figure 4.25). The models are very similar except for the derivation of the apparent dissociation constants (K_1, K_2, K_3, K_4) for each enzyme species.

Chapter 4   Experiment 4.10*

Figure 4.28

Formulas

	A	B	C	D	E	F
1	S(0.5) =	1	[S] =	0.5	1.0	2.0
2	ΔΔG(bnd) =	-1.4				
3			[E] =	Formula Set I		
4	alpha =	=10^(B2/1.4)	[ES] =	Formula Set II		
5	Kint =	=B1/(B4^1.5)	[ES2] =			
6			[ES3] =			
7	K1 =	=B5/4	[ES4] =			
8	K2 =	=2*B4*B5/3	Frtn Sites Bound =	Formula Set III		
9	K3 =	=3*B4*B4*B5/2				
10	K4 =	=4*B4*B4*B4*B5				

Values

	A	B	C	D	E	F
1	S(0.5) =	1	[S] =	0.5	1	2
2	ΔΔG(bnd) =	-1.4				
3			[E] =	0.8646	0.4323	0.0540
4	alpha =	0.10	[ES] =	0.0547	0.0547	0.0137
5	Kint =	31.62	[ES2] =	0.0130	0.0259	0.0130
6			[ES3] =	0.0137	0.0547	0.0547
7	K1 =	7.91	[ES4] =	0.0540	0.4323	0.8646
8	K2 =	2.11	Frtn Sites Bound =	0.0844	0.5000	0.9156
9	K3 =	0.47				
10	K4 =	0.13				

Formula Set I
Prototype Cell is D3
=1/(1+(D1/**B7**)+((D1*D1)/(**B7*B8**))+((D1*D1*D1)/(**B7*B8*B9**))
+((D1*D1*D1*D1)/(**B7*B8*B9*B10**)))

Formula Set II
Prototype Cell is D4
=(D3*D**1**)/**B**7

Formula Set III
Prototype Cell is D8
=(D4+(2*D5)+(3*D6)+(4*D7))/4

Variable Degree of Cooperativity

Questions

1. When the binding of a ligand molecule at one site does not affect the free energy of binding at other sites (i.e., when $\Delta\Delta G_{bind}$ is zero), the model described in Figure 4.28 should be equivalent to an enzyme with four independent catalytic sites (Experiment 4.9). Is it?

$S_{0.5} = 1 \quad \Delta\Delta G_{bind} = 0$

Fraction of Sites Filled

[S]	log [S]	[E]	[ES]	[ES$_2$]	[ES$_3$]	[ES$_4$]	Total Sites
0							
0.01	-2.0						
0.0316	-1.5						
0.1	-1.0						
0.316	-0.5						
1	0.0						
3.16	0.5						
10	1.0						
31.6	1.5						
100	2.0						

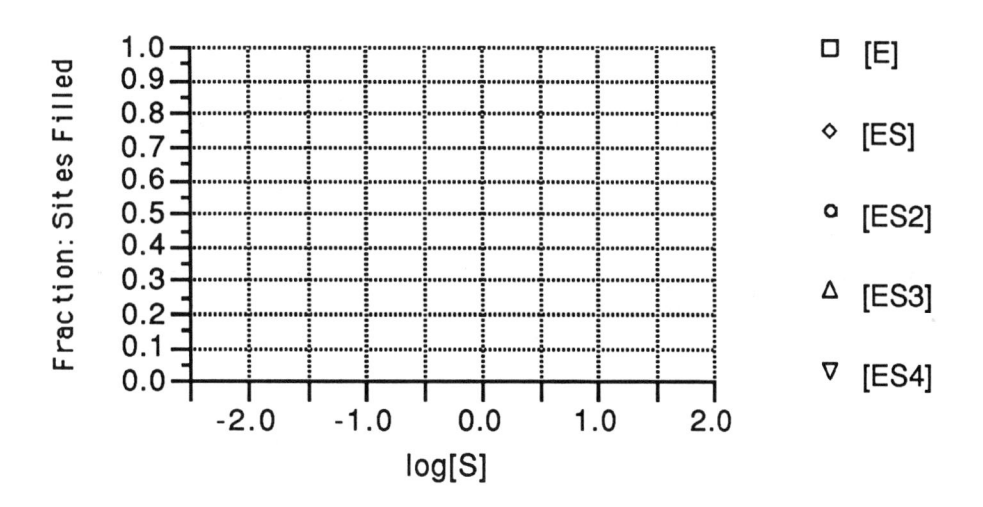

Chapter 4   Experiment 4.10*

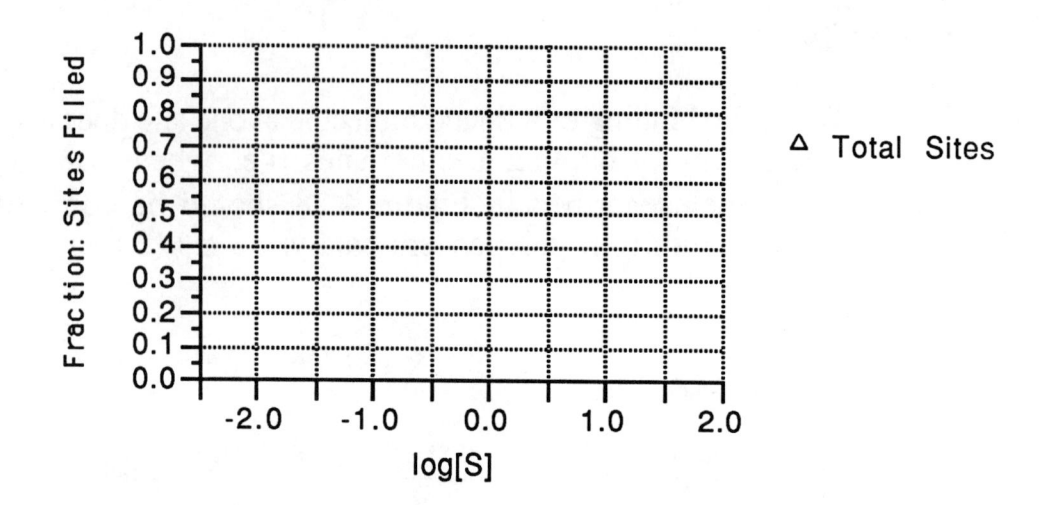

[Graph: Fraction: Sites Filled (y-axis, 0.0 to 1.0) vs log[S] (x-axis, -2.0 to 2.0); legend: △ Total Sites]

a. How do the data plotted above compare to the data plotted in Question 1 of Experiment 4.9?

2. When the binding of a ligand molecule at one site does affect the free energy of binding at other sites (i.e., when $\Delta\Delta G_{bind}$ is not zero), the model described in Figure 4.28 should be equivalent to an enzyme with four cooperative catalytic sites. Explore an enzyme with moderate cooperativity where $\Delta\Delta G_{bind}$ = -0.9 kcal/mol.

$$S_{0.5} = 1 \quad \Delta\Delta G_{bind} = -0.9$$

Fraction of Sites Filled

[S]	log [S]	[E]	[ES]	[ES$_2$]	[ES$_3$]	[ES$_4$]	Total Sites
0							
0.01	-2.0						
0.0316	-1.5						
0.1	-1.0						
0.316	-0.5						
1	0.0						
3.16	0.5						
10	1.0						
31.6	1.5						
100	2.0						

Variable Degree of Cooperativity

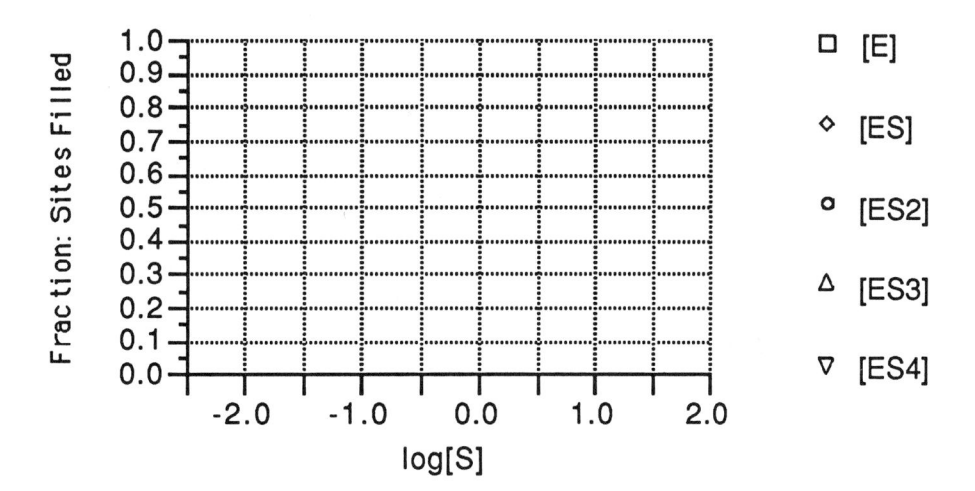

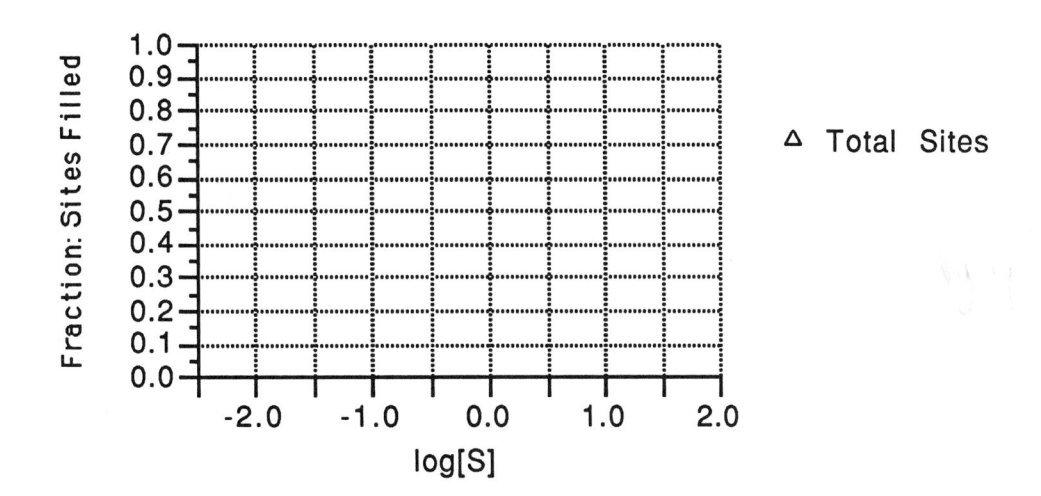

[Plot 1: Fraction: Sites Filled vs log[S], with legend: □ [E], ◇ [ES], ○ [ES2], △ [ES3], ▽ [ES4]]

[Plot 2: Fraction: Sites Filled vs log[S], with legend: △ Total Sites]

a. When the binding of ligand at one site increases the strength of binding at remaining sites (i.e., when $\Delta\Delta G_{bind}$ is negative), the probability that an enzyme will exist as ES_1, ES_2, or ES_3 should decrease. Does it?

b. On your own, explore a cooperative enzyme where $\Delta\Delta G_{bind}$ = -2.4 kcal/mol. As cooperativity increases, what happens to the concentrations of ES_1, ES_2, and ES_3 over a range of values of [S]?

3. Many electronic spreadsheet programs allow special display formats for cells that cause the cell display to simulate a histogram. Other electronic spreadsheet programs have more advanced, integrated graphics capabilities. If your electronic spreadsheet

Chapter 4   Experiment 4.10*

program supports such features, revise the model described in Figure 4.28 so that values of [E], [ES], [ES$_2$], [ES$_3$], and [ES$_4$] are displayed as histograms (e.g., as in Figure 4.29). Prepare histograms for a variety of combinations of [S] and $\Delta\Delta G_{bind}$, including

$S_{0.5} = 1$	$\Delta\Delta G_{bind} = 0$	(a noncooperative enzyme)
$S_{0.5} = 1$	$\Delta\Delta G_{bind} = -0.2$	
$S_{0.5} = 1$	$\Delta\Delta G_{bind} = -0.4$	

and so forth (up to a value of $\Delta\Delta G_{bind}$ where the enzyme is not distinguishable from an enzyme with infinite cooperativity [Experiment 4.7]).

Design your model so that as much as possible of your screen width can be used to display the histogram. You should multiply the contents of cells displaying as histograms by an assignable scaling factor to control the length of the bars (also, correct for this scaling factor by modifying, for example, the formulas in cells B14 through B17 in Figure 4.29).

Figure 4.29

Formulas

	A	B
1	S(0.5) =	1
2	$\Delta\Delta G$(bind) =	-1.4
3	[S] =	0.5
4	Scaling Factor =	40
5		
6	alpha =	=10^(B2/1.4)
7	Kint =	=B1/(B6^1.5)
8	K1 =	=B7/4
9	K2 =	=2*B6*B7/3
10	K3 =	=3*B6*B6*B7/2
11	K4 =	=4*B6*B6*B6*B7
12	[E] =	=B4/(1+(B3/B8)+((B3*B3)/(B8*B9))+((B3*B3*B3)/(B8*B9*B10))+((B3*B3*B3*B3)/(B8*B9*B10*B11)))
13	[ES] =	=(((B12/B4)*B3)/B8)*B4
14	[ES2] =	=(((B13/B4)*B3)/B9)*B4
15	[ES3] =	=(((B14/B4)*B3)/B10)*B4
16	[ES4] =	=(((B15/B4)*B3)/B11)*B4
17	Frtn Sites Bound =	=((B13/B4)+(2*(B14/B4))+(3*(B15/B4))+(4*(B16/B4)))/4

(Continued.)

Variable Degree of Cooperativity

Figure 4.29 (Contd.)

Values

	A	B
1	S(0.5) =	1
2	ΔΔG(bind) =	-1.4
3	[S] =	0.5
4	Scaling Factor =	40
5		
6	alpha =	0.10
7	Kint =	31.62
8	K1 =	7.91
9	K2 =	2.11
10	K3 =	0.47
11	K4 =	0.13
12	[E] =	********************************
13	[ES] =	**
14	[ES2] =	*
15	[ES3] =	*
16	[ES4] =	**
17	Frtn Sites Bound =	0.08

a. In this new format, reexamine the issues raised by Question 2. What happens to the distribution of enzyme forms (E, ES_1, ES_2, etc.) as $\Delta\Delta G_{bind}$ increases in magnitude (i.e., negative values increase)? At approximately what point does the enzyme modeled in Figure 4.29 become indistinguishable from the infinitely cooperative enzyme described in Experiment 4.7?

Recall that the strength of a single hydrogen bond is typically around 3 kcal/mol. Although it is not strictly correct to compare bond strengths (ΔH) to binding energies (ΔG), the comparison is valid as an illustration of how small a change in binding energy is required to produce effectively infinite cooperativity.

b. How do changes in the value of $[S]/S_{0.5}$ influence the distribution of enzyme forms at different values of $\Delta\Delta G_{bind}$?

c. An enzyme that displays even relatively modest cooperativity is distributed almost totally between E and ES_4 at all concentrations of substrate. It is this fact that leads to fourth-order kinetics for such enzymes. If ES_4 is the only reactive species to occur in significant concentration, the rate of the reaction is equal to $k_{cat}[ES_4]$.

Chapter 4   Experiment 4.10*

At low concentrations of S when most sites are unfilled, the concentration of ES_4, and thus the velocity of the reaction, is proportional to $[S]^4$. Why is the reaction not of fourth order at higher concentrations of S?

Problems

The histogram model described in Question 2 offers you a powerful tool into the behavior of cooperative enzymes. Many insights to be obtained from the model come from the instantaneous feedback that derives from varying a parameter, observing the results, and asking new questions that arise as a consequence of your results.

You may wish to take the time to explore this model by varying its assignable parameters singly and in groups while simply watching your screen. As your explorations raise questions, use your model to answer them.

Experiment 4.11*

Cooperative Kinetics: The Hill Equation

The rates of first-order, homogeneous chemical reactions typically are proportional to the concentration of reactant even at very high concentrations. As the concentration of reactant increases, the reaction velocity increases linearly. In Experiment 4.3 you saw that, for noncooperative enzymic reactions, the value of $v/(V_{max} - v)$ increases linearly with the concentration of reactant. The derived quantity $v/(V_{max} - v)$ behaves (in noncooperative enzymic reactions) similarly to v (in first-order, homogeneous chemical reactions) because $v/(V_{max} - v)$ "strips out" the effects of catalytic-site saturation (examine Equation 4.8). In this experiment you will discover that this important property of $v/(V_{max} - v)$ may be generalized to all enzymic reactions.

The velocity of any enzyme-catalyzed reaction must asymptotically approach a maximal value because of catalytic site saturation (Figure 4.30a and b). However, whereas the slope of the plot of $v/(V_{max} - v)$ vs. [S] is a constant for noncooperative enzymes (Figure 4.30c), the slope of the plot of $v/(V_{max} - v)$ vs. [S] for cooperative enzymes increases with [S] (Figure 4.30d). In fact, the curves in Figures 4.30c and d are identical to plots of v vs. [S] for homogeneous reactions of first order and higher order, respectively.

Just as the kinetic order of a *homogeneous* reaction is equal to the slope of the plot of log velocity vs. log reactant concentration, one useful measure of the kinetic order of an *enzymic* reaction is the slope of the plot of log $v/(V_{max} - v)$ vs. log[S] Figure 4.30e. This relationship was first used by A.V. Hill in his studies of the binding of oxygen to hemoglobin. The plot of log $[v/(V_{max} - v)]$ vs. log[S] and the equation for such a plot are known as the Hill plot and Hill equation, respectively.

Chapter 4   Experiment 4.11*

Figure 4.30

Noncooperative Reactions   **Cooperative Reactions**

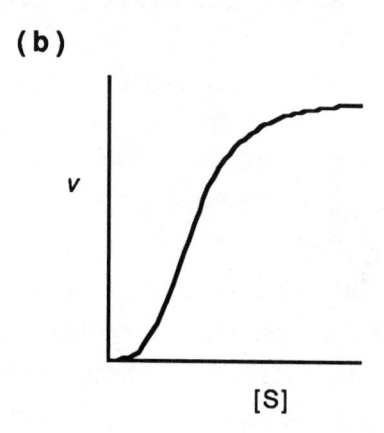

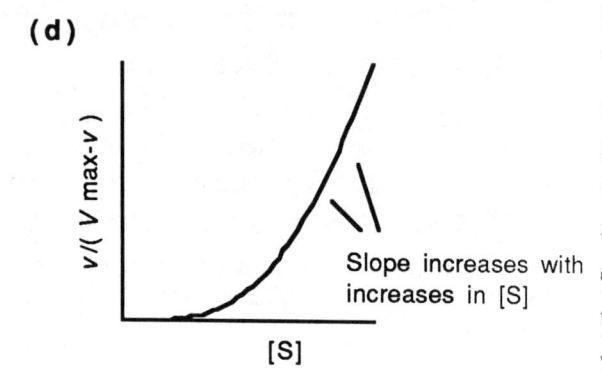

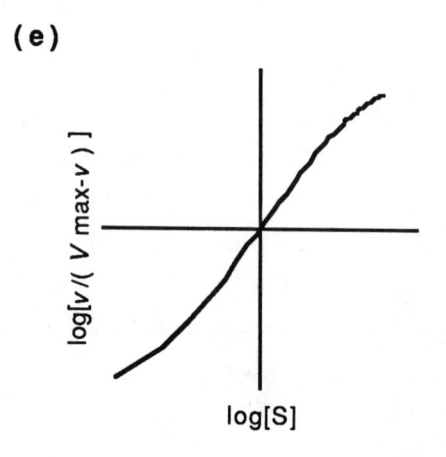

(a) v vs. [S]

(b) v vs. [S]

(c) $v/(V_{max}-v)$ vs. [S] — Slope constant for all values of [S]

(d) $v/(V_{max}-v)$ vs. [S] — Slope increases with increases in [S]

(a) and (b) are plots of v vs. [S] for noncooperative and cooperative enzymic reactions. Such plots plateau at high values of [S] because of catalytic site saturation.

(c) and (d) are plots of $v/(V_{max}-v)$ vs. [S]. $v/(V_{max}-v)$ "strips out" the effects of catalytic site saturation.

(e) is a Hill plot. The slope of Hill plots is one measure of reaction order.

(e) $\log[v/(V_{max}-v)]$ vs. $\log[S]$

142

The Hill Equation

Of course, the true kinetic order of an enzymic reaction must be defined the same way as the kinetic order of any other chemical reaction (i.e., the slope of a plot of log *v* vs. log [S]). At low values of [S], the true kinetic order and kinetic order as defined by the Hill equation are nearly identical. The two different calculations yield different results as [S] increases and catalytic sites become filled.

The Hill order may be regarded as the intrinsic kinetic order of the reaction — the order after the effects of site saturation have been removed. For a simple, noncooperative Michaelis reaction the Hill order is one for all concentrations of substrate. For this reason such reactions are usually considered to be of first order with respect to [S].

Directions

1. Build and verify the model described in Figure 4.31.

2. The model is very similar to that described in Figure 4.28. Values are calculated for closely adjacent values of [S] to determine the value of the slopes of tangent lines required to determine the true kinetic order and the Hill order.

Figure 4.31

Formulas

	A	B	C	D	E
1				[S]	[S] + Δ[S]
2	S(0.5) =	1		1	=1.001*D2
3	ΔΔG(bind) =	-1.4			
4			[E] =	Formula Set I	
5	alpha =	=10^(B3/1.4)	[ES] =	Formula Set II	
6	Kint =	=B2/(B5^1.5)	[ES2] =		
7			[ES3] =		
8	K1 =	=B6/4	[ES4] =		
9	K2 =	=2*B5*B6/3	Fraction Sites Bound =	Formula Set III	
10	K3 =	=3*A5*A5*A6/2			
11	K4 =	=4*B5*B5*B5*B6	v/Vmax =	=D9	=E9
12			v/(Vmax - v) =	=D11/(1-D11)	=E11/(1-E11)
13					
14			Hill Slope =	=(LOG(E12)-LOG(D12))/(LOG(E2)-LOG(D2))	
15			Momentary Rxn Order =	=(LOG(E11)-LOG(D11))/(LOG(E2)-LOG(D2))	

(Continued.)

Chapter 4 **Experiment 4.11***

dyn144.wk1

Figure 4.31 (Contd.)

Values

	A	B	C	D	E
1				[S]	[S] + Δ[S]
2	S(0.5) =	1		1	1.001
3	ΔΔG(bind) =	-1.4			
4			[E] =	0.4323	0.4315
5	alpha =	0.10	[ES] =	0.0547	0.0546
6	Kint =	31.62	[ES2] =	0.0259	0.0259
7			[ES3] =	0.0547	0.0547
8	K1 =	7.91	[ES4] =	0.4323	0.4332
9	K2 =	2.11	Fraction Sites Bound =	0.5000	0.5009
10	K3 =	0.47			
11	K4 =	0.13	v/Vmax =	0.5000	0.5009
12			v/(Vmax - v) =	1.0000	1.0036
13					
14			Hill Slope =	3.5681	
15			Momentary Rxn Order =	1.7825	

Formula Set I
Prototype Cell is D4
=1/(1+(D2/**B8**)+((D2*D2)/(**B8*B9**))+((D2*D2*D2)/(**B8*B9*B10**))
+((D2*D2*D2*D2)/(**B8*B9*B10*B11**)))

Formula Set II
Prototype Cell is D5
=(D4*D**2**)/**B**8

Formula Set III
Prototype Cell is D9
=(D5+(2*D6)+(3*D7)+(4*D8))/4

144

The Hill Equation

Questions

1. Use the model described in Figure 4.31 to complete and graph the table below by finding the true kinetic order and the kinetic order as determined by the Hill equation.

$\Delta\Delta G_{bind} =$	0		-0.9		-1.4		-2.8 (kcal/mol)	
[S]	Hill	True	Hill	True	Hill	True	Hill	True
0.010	___	___	___	___	___	___	___	___
0.030	___	___	___	___	___	___	___	___
0.100	___	___	___	___	___	___	___	___
0.200	___	___	___	___	___	___	___	___
0.333	___	___	___	___	___	___	___	___
0.500	___	___	___	___	___	___	___	___
1.000	___	___	___	___	___	___	___	___
1.500	___	___	___	___	___	___	___	___
2.000	___	___	___	___	___	___	___	___
3.000	___	___	___	___	___	___	___	___
5.000	___	___	___	___	___	___	___	___
10.000	___	___	___	___	___	___	___	___
100.000	___	___	___	___	___	___	___	___

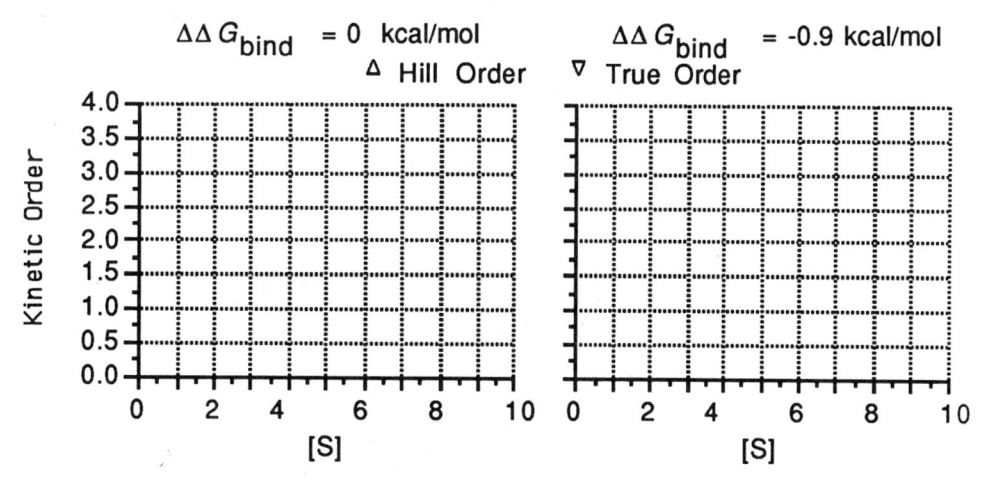

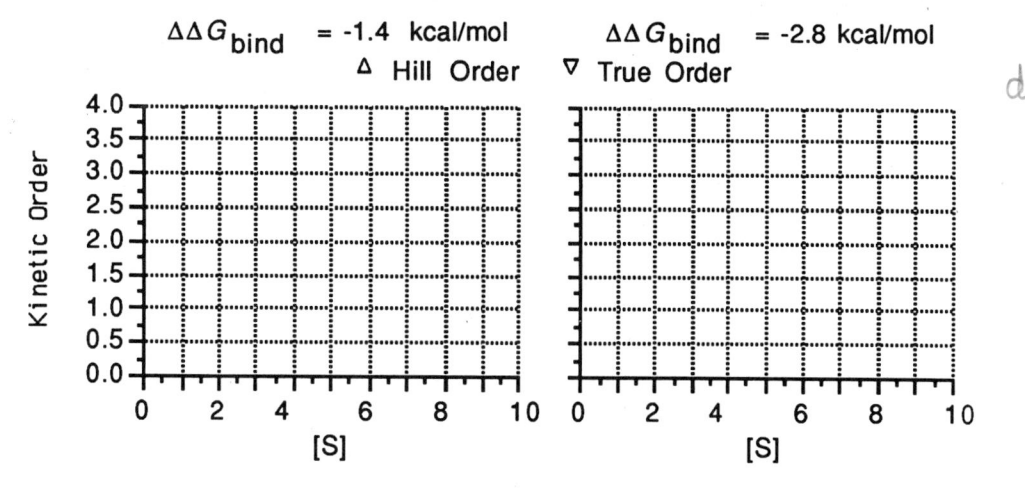

$\Delta\Delta G_{bind}$ = 0 kcal/mol
△ Hill Order   ▽ True Order

$\Delta\Delta G_{bind}$ = -0.9 kcal/mol
△ Hill Order   ▽ True Order

$\Delta\Delta G_{bind}$ = -1.4 kcal/mol
△ Hill Order   ▽ True Order

$\Delta\Delta G_{bind}$ = -2.8 kcal/mol
△ Hill Order   ▽ True Order

Chapter 4   Experiment 4.11*

a. Examine the true order and Hill order of a noncooperative enzyme ($\Delta\Delta G_{bind} = 0$). You should recognize the behavior from your exploration of Experiment 4.3. The Hill order has a value of one at all values of [S]. When the effects of site saturation are not "stripped out," true kinetic order varies from a value of one (at low values of [S]) to zero (at high values of [S]).

b. Examine your data for cooperative enzymes. The Hill order approaches a value of one at both very high and very low values of [S].

At low values of [S], the only enzyme species present in significant amounts are E and ES (ES will almost always convert to EP before ES has an opportunity to bind additional S). Kinetic behavior at low values of [S] will therefore resemble that of a first-order enzyme.

At high values of [S], the only enzyme species present in significant amounts are ES_3 and ES_4 (ES_3 will almost always bind S to its empty site before any of the other three, filled catalytic sites have time to react). Kinetic behavior at high values of [S] will therefore also resemble that of a single-site enzyme.

c. At what value of [S] does the Hill order assume its maximum value? (Answer for several values of $S_{0.5}$.)

d. Compare your Hill-order data for cooperative enzymes to the values obtained for true kinetic order.

At low values of [S], Hill order is indistinguishable from true order. As [S] increases, Hill order and true order have different values. Why?

The Hill Equation

2. Complete the table below.

	$S_{0.5} = 1$ $\Delta\Delta G_{bind} = -1.4$ kcal/mol		$S_{0.5} = 1$ $\Delta\Delta G_{bind} = -2.8$ kcal/mol	
[S]	Hill Order	True Order	Hill Order	True Order
0.500	_____	_____	_____	_____
1.000	_____	_____	_____	_____
2.000	_____	_____	_____	_____
0.333	_____	_____	_____	_____
1.000	_____	_____	_____	_____
3.000	_____	_____	_____	_____
0.200	_____	_____	_____	_____
1.000	_____	_____	_____	_____
5.000	_____	_____	_____	_____

Notice that values of the Hill order not only peak when $[S] = S_{0.5}$, but also are symmetrical about $S_{0.5}$. Flesh out the table with additional values to confirm this point, if you wish.

Values of the true order, by contrast, peak at a value of [S] that varies with $\Delta\Delta G_{bind}$ and then decrease steadily toward a value of 0 (as the true orders of all enzyme-catalyzed reactions must).

3. Use the table developed in Question 1 to answer the following questions.

   a. How does the maximal value of the Hill order change as $\Delta\Delta G_{bind}$ increases in magnitude from 0 to -5?

   b. How does the rate at which the Hill order falls off on each side of the midpoint change with changes in $\Delta\Delta G_{bind}$?

   As $\Delta\Delta G_{bind}$ assumes large negative values, the model approaches the case of infinite cooperativity explored in Experiment 4.7. Verify that, at infinite cooperativity, the value of the Hill order corresponds to the number of catalytic sites at all values of [S].

Chapter 4   Experiment 4.11*

Real cooperative enzymes (e.g., $\Delta\Delta G_{bind}$ = -2.8 kcal/mol in the preceding table) often appear to be infinitely cooperative (Hill order = number of catalytic sites) over a wide range around $S_{0.5}$. No enzyme can be truly infinitely cooperative, however, as can be seen by entering very low or very high values of [S] into your model.

Problems

1. The model described in Figure 4.31 approximates the behavior of oxygen-hemoglobin binding, for which $\Delta\Delta G_{bind} \sim$ -0.9 kcal/mol. Each hemoglobin molecule contains four heme groups, and the order of binding is about 2.7. Verify that small changes in $\Delta\Delta G_{bind}$ would dramatically change the kinetics of oxygen-hemoglobin binding. Note how small a change in binding energy underlies the cooperative pattern of binding in this important case.

2. The model described in Figure 4.31 may be changed in a number of ways to explore a wider range of substrate-enzyme interactions. For example, for some cooperative enzymes and other proteins that bind ligand cooperatively, it seems that the value of α is not constant.

    Set up a model in which the interaction factors of the first, second, and third bound ligand are independently assignable. Explore the pattern of Hill slopes that result if most of the change in binding affinity has already occurred when the first molecule of ligand has bound (and there is little further change as ligand occupies additional sites). This situation corresponds to the Monod-Wyman-Changeux model that is discussed in most textbooks. Notice that when α is not constant, three of the properties that we observed for the simple case are no longer true: The maximal slope is not at the midpoint; the curve is not symmetrical around the midpoint; and Equation 4.49 ($K_{int} = S_{0.5}/\alpha^{1.5}$) is not valid. Thus in such models the assignable parameter should be K_{int} rather than $S_{0.5}$.

    Set up a model in which there is little or no change in the strength of binding until the third site is filled.

Experiment 4.12*

Cooperative Kinetics: Modifier Effects

The typical control element in metabolism is an enzyme that bears two or more interacting catalytic sites and two or more interacting modifier sites. The binding of substrates and of modifiers is usually positively cooperative, and the interactions between modifier and substrates may be either positive or negative, whichever is appropriate for regulation of each enzyme.

In this experiment you will build a model to simulate the simplest such enzyme — an enzyme with two catalytic sites and two modifier sites.

Directions

1. Build and verify the model described in Figure 4.32.

2. If at all possible, you should format the cells D4 through D12 to display a histogram rather than numbers. The model is arranged to allow the greatest possible screen width for histogram display. The calculations below row 18 need not appear on your screen.

3. The assignable parameters are:

   [S], [M]: The concentrations of substrate and modifier

   $K_{int(s)}$: The intrinsic dissociation constant for substrate

   $K_{int(m)}$: The intrinsic dissociation constant for modifier

   $\Delta\Delta G_s$: The change in free energy at a second catalytic site associated with substrate binding at a first catalytic site

   $\Delta\Delta G_m$: The change in free energy at a second modifier site associated with modifier binding at a first modifier site

   $\Delta\Delta G_{ms}$: The change in free energy at a catalytic site associated with the binding of modifier binding (or the change in free energy at a modifier site associated with binding of substrate)

4. The mathematics of the model is very similar to that already developed and will be left as an exercise for the student.

Chapter 4   Experiment 4.12*

Figure 4.32

Formulas

	A	B
1	0.1	= [S]
2	0.2	= [M]
3	0.15	= Kint(s)
4	0.2	= Kint(m)
5	-1.4	= ΔΔGs
6	-1.2	= ΔΔGm
7	1.4	=ΔΔGms
8	2	% of Etot equals 1 unit
9	E	=(100*A37)/A8
10	M-E	=(100*A38)/A8
11	2M-E	=(100*A39)/A8
12	E-S	=(100*A40)/A8
13	M-E-S	=(100*A41)/A8
14	2M-E-S	=(100*A42)/A8
15	E-2S	=(100*A43)/A8
16	M-E-2S	=(100*A44)/A8
17	2M-E-2S	=(100*A45)/A8
18	v/Vmax =	=A8*(B12+B13+B14+(2*(B15+B16+B17)))/200
19		
20	=10^(A5/1.4)	= F(S)
21	=10^(A6/1.4)	=F(M)
22	=10^(A7/1.4)	= F(MS)
23	=A3/2	= Kint(s)1
24	=2*A3*A20	= Kint(s)2
25	=A4/2	= Kint(m)1
26	=2*A4*A21	= Kint(m)2
27	1	E
28	=A2/A25	M-E
29	=(A2*A2)/(A25*A26)	2M-E
30	=A1/A23	E-S
31	=(A1*A2)/(A23*A25*A22)	M-E-S
32	=(A2*A2*A1)/(A26*A25*A23*A22*A22)	2M-E-S
33	=(A1*A1)/(A23*A24)	E-2S
34	=(A2*A1*A1)/(A25*A24*A23*A22*A22)	M-E-2S
35	=(A1*A1*A2*A2)/(A23*A24*A25*A26*A22*A22*A22*A22)	2M-E-2S
36	=SUM(A27..A35)	Etot
37	=1/A36	E/Etot
38	=A37*A28	M-E/Etot
39	=A37*A29	2M-E/Etot
40	=A37*A30	E-S/Etot
41	=A37*A31	M-E-S/Etot
42	=A37*A32	2M-E-S/Etot
43	=A37*A33	E-2S/Etot
44	=A37*A34	M-E-2S/Etot
45	=A37*A35	2M-E-2S/Etot

(Continued.)

Figure 4.32 (Contd.)

Values

	A	B
1	0.1	= [S]
2	0.2	= [M]
3	0.15	= Kint(s)
4	0.2	= Kint(m)
5	-1.4	= ΔΔGs
6	-1.2	= ΔΔGm
7	1.4	= ΔΔGms
8	2	% of Etot equals 1 unit
9	E	3.04
10	M-E	6.09
11	2M-E	21.90
12	E-S	4.06
13	M-E-S	0.81
14	2M-E-S	0.29
15	E-2S	13.53
16	M-E-2S	0
17	2M-E-2S	0
18	v/Vmax =	0.3277
19		
20	0.1	= F(S)
21	0.1389	=F(M)
22	10	= F(MS)
23	0.075	= Kint(s)1
24	0.03	= Kint(s)2
25	0.1	= Kint(m)1
26	0.05558	= Kint(m)2
27	1	E
28	2.0000	M-E
29	7.1969	2M-E
30	1.3333	E-S
31	0.2667	M-E-S
32	0.0960	2M-E-S
33	4.4444	E-2S
34	0.0889	M-E-2S
35	0.0032	2M-E-2S
36	16.4293	Etot
37	0.0609	E/Etot
38	0.1217	M-E/Etot
39	0.4380	2M-E/Etot
40	0.0812	E-S/Etot
41	0.0162	M-E-S/Etot
42	0.0058	2M-E-S/Etot
43	0.2705	E-2S/Etot
44	0.0054	M-E-2S/Etot
45	0.0002	2M-E-2S/Etot

(Display these cells as histograms rather than numbers, if possible)

Recalculation Order = Columnwise

Chapter 4   Experiment 4.12*

Questions

1. Study the model described in Figure 4.32 and derive the equations used to build the model.

    For all of the questions that follow you will find it helpful to build your answers on your computer screen (using the histogram format). Otherwise record your answers on a form such as that displayed below.

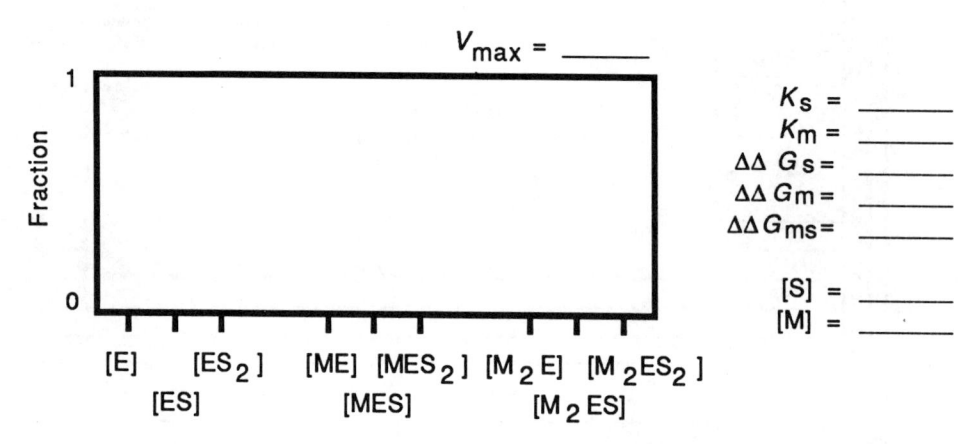

    V_{max} = _____

    K_s = _____
    K_m = _____
    $\Delta\Delta G_s$ = _____
    $\Delta\Delta G_m$ = _____
    $\Delta\Delta G_{ms}$ = _____

    [S] = _____
    [M] = _____

    [E] [ES] [ES$_2$] [ME] [MES] [MES$_2$] [M$_2$E] [M$_2$ES] [M$_2$ES$_2$]

2. Set up the assignable parameters of the model as follows:

$K_{int(s)}$ = 1		$\Delta\Delta G_s$ = 0		[S] = *	
$K_{int(m)}$ = 1		$\Delta\Delta G_m$ = 0		[M] = 0	
		$\Delta\Delta G_{ms}$ = 0			

    *Vary [S] from 0 to 10.

    This is the behavior of a simple Michaelis enzyme. Remember it for comparative purposes.

3. Set up the assignable parameters of the model as follows:

$K_{int(s)}$ = 1		$\Delta\Delta G_s$ = -1.4		[S] = *	
$K_{int(m)}$ = 1		$\Delta\Delta G_m$ = 0		[M] = 0	
		$\Delta\Delta G_{ms}$ = 0			

    *Vary [S] from 0 to 10.

    This is the simplest type of cooperative enzyme — with two interacting sites and no modifier effect. Notice how small changes in the value of [S] affect larger changes in the distribution of enzyme-substrate complexes than in the case of noncooperative enzymes. What changes do you see in the *pattern* of enzyme-substrate complex distribution?

4. Set up the assignable parameters of the model as follows:

$K_{int(s)} = 1$ $\Delta\Delta G_s = *$ $[S] = 5$
$K_{int(m)} = 1$ $\Delta\Delta G_m = 0$ $[M] = 0$
              $\Delta\Delta G_{ms} = 0$

*Vary $\Delta\Delta G_s$ between 2.8 and -2.8.

This is also the simplest type of cooperative enzyme. Observe the ratio of [ES] and [ES$_2$] at varying values of $\Delta\Delta G_s$: Negative cooperativity, as illustrated by positive values of $\Delta\Delta G_s$, may occur in nature but little is known of its occurrence or function. How do the binding patterns of positively and negatively cooperative enzymes differ?

5. Set up the assignable parameters of the model as follows:

$K_{int(s)} = 1$ $\Delta\Delta G_s = -1.4$ $[S] = $ Any value
$K_{int(m)} = 1$ $\Delta\Delta G_m = -1.4$ $[M] = *$
              $\Delta\Delta G_{ms} = 0$

* Vary [M] over any arbitrary range and notice how sensitive the distribution of enzyme species is to changes in [M].

v/V_{max} is not altered by changes in [M] when $\Delta\Delta G_{ms} = 0$. Explain.

6. Set up the assignable parameters of the model as follows:

$K_{int(s)} = 1$ $\Delta\Delta G_s = -1.4$ $[S] = *$
$K_{int(m)} = 1$ $\Delta\Delta G_m = -1.4$ $[M] = **$
              $\Delta\Delta G_{ms} = -1.4$

* Set [S] such that v/V_{max} falls between 0.05 and 0.10 in the absence of modifier.

** Increase [M] in small steps beginning at [M] = 0 and notice how sensitive the reaction rate is to changes in [M].

Chapter 4   Experiment 4.12*

Notice that even with this very small effect on binding energies and with only two sites of each kind, the enzyme partitions mainly between E and M_2ES_2. Change the $\Delta\Delta G$ values to -2.8 and observe the sharpening of the cooperative response.

7. Set up the assignable parameters of the model as follows:

$K_{int(s)}$	= 1	$\Delta\Delta G_s$	= -1.4	[S]	= *
$K_{int(m)}$	= 1	$\Delta\Delta G_m$	= -1.4	[M]	= **
		$\Delta\Delta G_{ms}$	= 1.4		

* Set [S] such that v/V_{max} falls between 0.5 and 0.7 in the absence of modifier.

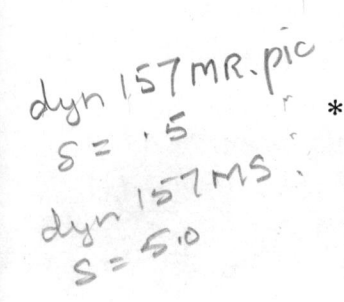

** Increase [M] in small steps beginning at [M] = 0.

These parameter settings simulate what is probably the most common type of regulatory enzyme — an enzyme with positively cooperative binding of both substrate and modifier and negative interactions between substrate and modifier.

Most biosynthetic pathways are regulated in part by such interactions at the enzyme that catalyzes the first step in the sequence. The effects are usually stronger in nature than in this simulation because:

a. There are frequently more than two interacting catalytic sites (all else being equal, the strength of the regulatory effect is squared if the number of binding sites increases from two to four).

b. The change in binding energy is usually greater than 1.4 kcal/mol.

8. Systematic variation of the parameters of this model illustrates numerous features of enzymes of this important type.

   A few suggestions:

   a. Set up the assignable parameters of the model as follows:

$K_{int(s)} = 0.15$	$\Delta\Delta G_s = -1.4$	$[S] = 0.1$
$K_{int(m)} = 0.20$	$\Delta\Delta G_m = -1.4$	$[M] = 0.1$
	$\Delta\Delta G_{ms} = *$	

   *Vary $\Delta\Delta G_{ms}$ from 2.8 to -2.8 kcal/mol.

   b. Set all $\Delta\Delta G$ terms at large positive values (e.g., 10 kcal/mol). The results should generate a "competitive" type of Lineweaver-Burk plot that could not be distinguished from the very different case of competition of S and M for a single site. Can you explain this?

Problems

1. Modify the model described in Figure 4.32 so that you can monitor the rate of change of v/V_{max} with respect to [M] (e.g., set up an additional column of calculations that compute v/V_{max} at a closely neighboring point — a slightly different value of [M] — and use the results to calculate the slope of the tangent line between the two points).

   Create two different enzyme behaviors: (1) an enzyme that is positively cooperative with respect to binding of substrate and modifier and *positively* cooperative with respect to the interactions between substrate and modifier, and (2) an enzyme that is positively cooperative with respect to binding of substrate and modifier and *negatively* cooperative with respect to the interactions between substrate and modifier.

   Find values of [S] for each such that [M] can have a significant impact on v/V_{max} and determine the range of values of [M] that have a maximal impact on the rate of change of v/V_{max}.

Chapter 4   Experiment 4.12*

You should observe that the region of maximal sensitivity to modifier concentration is at a higher concentration for an enzyme that is negatively cooperative with respect to the interactions between substrate and modifier (relative to the value of K_m) than for an enzyme that is positively cooperative with respect to the interactions between substrate and modifier. Why?

Experiment 4.13*

Cooperative Kinetics: Stimulation and Inhibition of Reaction Rates by Substrate Analogs

Substrate analogs compete with substrate at catalytic sites but do not react to form product. You explored the effect of such analogs on the behavior of first order enzymes in Experiment 4.5. In this experiment you will explore the effect of such analogs on cooperative enzymes.

Under the terms of the Michaelis treatment for single site enzymes, the concentration of ligand-bound enzyme is proportional to ligand concentration divided by its dissociation constant (Equations 4.6 and 4.50).

$$\frac{[S]}{K_s} = \frac{[ES]}{[E]}$$

$$\frac{[A]}{K_a} = \frac{[EA]}{[E]} \qquad (4.50)$$

where S = substrate
A = substrate analog
K_s = dissociation constant for substrate
K_a = dissociation constant for analog

In consequence, substrate analogs have relatively little effect on reaction rate when both substrate and analog concentrations are low and most of the enzyme is free. On the other hand, when most enzyme molecules are bound to either substrate or analog, reaction rate is proportional to the ratio of $[S]/K_s$ to $[A]/K_a$ and is therefore independent of absolute concentrations. You should take a moment to verify these statements either intuitively or algebraically.

These same relationships apply for enzymes that bind substrate (and substrate analogs) cooperatively, but the kinetic consequences can be very different.

Chapter 4   Experiment 4.13*

1. At high ligand concentrations, when there is little free enzyme, reaction rate is proportional to the ratio of $[S]/K_s$ to $[A]/K_a$ — just as in the case of a single-site enzyme.

2. At low ligand concentrations, reaction rates become a function of the manner in which a molecule of bound analog affects the affinities of substrate and analog at other sites.

You will explore these relationships in this experiment. The model illustrates cases where analog bound at a catalytic site increases the affinity of both analog and substrate at other sites. Examples of real enzyme systems that exhibit such behavior are aspartate transcarbamylase (succinate acts as an analog of aspartate) and mitochondrial isocitrate dehydrogenase (citrate acts as an analog of isocitrate).

For simplicity, the model mimics an enzyme with infinite cooperativity (Experiment 4.7). This is essentially true for yeast mitochondrial isocitrate dehydrogenase, where the reaction retains its high order at very low concentrations of substrate.

Directions

1. Build and verify the models described in Figures 4.12 (Experiment 4.5), 4.33, and 4.34.

2. The models described in Figures 4.33 and 4.34 use the mathematics described in Experiment 4.7 except that a factor has been added to account for analog that is bound to a catalytic site but does not contribute to the reaction rate (Equation 4.51)

$$\frac{v}{V_{max}} = \left(\frac{[S+A]^n}{[S_{0.5}]^n + [S+A]^n}\right)\left(\frac{[S]}{[S]+[A]}\right) \qquad (4.51)$$

where   S = substrate
         A = substrate analog
         n = number of sites
         $S_{0.5}$ = substrate concentration at which half of the catalytic sites are filled

3. The model assumes that the number of sites $n = 4$ and that the dissociation constants for substrate and analog are equal.

Figure 4.33

Formulas

	A	B
1	S(0.5) =	0.1
2	[Substrate] =	0.5
3	[Analog]/[Substrate] =	0.4
4		
5	[Analog] =	=B2*B3
6	v =	=(((B2+B5)^4)/(((B2+B5)^4)+(B1*B1*B1*B1)))*(B2/(B2+B5))

Values

	A	B
1	S(0.5) =	0.1
2	[Substrate] =	0.5
3	[Analog]/[Substrate] =	0.4
4		
5	[Analog] =	0.2
6	v =	0.71

Chapter 4   **Experiment 4.13***

Figure 4.34

dyn160.wk1

dyn160 m

Formulas

	A	B	C	D	E	F	G
1							% Effect of
2				[Analog]	v	v'	[Analog]
3	S(0.5) =	0.1		0	***		
4	[Substrate] =	0.04		0.005	=E3		
5				0.01	=E3	Formula Set I	Formula Set II
6				0.03	=E3		
7				0.05	=E3		
8				0.1	=E3		
9				0.3	=E3		
10				0.5	=E3		
11				1	=E3		
12				3	=E3		
13				5	=E3		
14				10	=E3		

Values

	A	B	C	D	E	F	G
1							% Effect of
2				[Analog]	v	v'	[Analog]
3	S(0.5) =	0.1		0	0.0250	0.0250	
4	[Substrate] =	0.04		0.005	0.0250	0.0350	40.3
5				0.01	0.0250	0.0471	88.5
6				0.03	0.0250	0.1106	343.2
7				0.05	0.0250	0.1761	605.4
8				0.1	0.0250	0.2267	808.2
9				0.3	0.0250	0.1168	367.8
10				0.5	0.0250	0.0740	196.4
11				1	0.0250	0.0385	54.1
12				3	0.0250	0.0132	-47.3
13				5	0.0250	0.0079	-68.2
14				10	0.0250	0.0040	-84.0

Cell E3 is =((B4*B4*B4*B4)/((B4*B4*B4*B4)+(B3*B3*B3*B3)))

Formula Set I
Prototype Cell is F3
=(((B4+D3)^4)/(((B4+D3)^4)+(B3*B3*B3*B3)))*(B4/(B4+D3))

Formula Set II
Prototype Cell is G4
=100*(F4-E4)/E3

Substrate Analogs

Questions

1. Refresh your knowledge of the features of competitive inhibition in single-site enzymes by using the model described in Figure 4.12 (Experiment 4.5) to complete the tables below:

   a. When substrate and competitive inhibitor are present in equal concentrations, is the inhibitor most effective at high or low concentrations of total ligand? Why? (Assume that no product is present and that the other assignable parameters are as displayed in Figure 4.12.)

[S]	[I]	% Inhibition
0.001	0.001	_____
0.01	0.01	_____
0.1	0.1	_____
1	1	_____
10	10	_____

   b. Rewrite the model described in Figure 4.12 so that the concentration of the competitive inhibitor is always a fixed proportion of the concentration of substrate (e.g., cell B3 contains = 0.25*B1 or some other factor). The concentration of substrate S is identified as [A] in Figure 4.12. Assume that no product is present.

   Do you understand why, as total ligand concentration increases, reaction rate becomes a function of the [I]/[S] ratio and independent of absolute concentrations of ligand?

   Reaction Rate

[S]	[I] = [S]/4	[I] = [S]	[I] = 4 * [S]
0.025	_____	_____	_____
0.05	_____	_____	_____
0.1	_____	_____	_____
0.25	_____	_____	_____
0.5	_____	_____	_____
1.0	_____	_____	_____
2.5	_____	_____	_____
5.0	_____	_____	_____
10.0	_____	_____	_____
25.0	_____	_____	_____
50.0	_____	_____	_____
100.0	_____	_____	_____

Chapter 4   Experiment 4.13*

2. Use the model described in Figure 4.33 to complete the table below and explore the effect of total ligand concentration on the reaction rate for a cooperative enzyme. Assume that $S_{0.5} = 0.2$. Notice that the substrate analog is now identified as A (analog) rather than I (inhibitor). The reason that this is appropriate will soon become apparent.

[S]	[A] = [S]/4	Reaction Rate [A] = [S]	[A] = 4 * [S]
0.025	___	___	___
0.05	___	___	___
0.1	___	___	___
0.25	___	___	___
0.5	___	___	___
1.0	___	___	___
2.5	___	___	___
5.0	___	___	___
10.0	___	___	___
25.0	___	___	___
50.0	___	___	___
100.0	___	___	___

a. As with noncooperative enzymes, reaction rate becomes a function of the [A]/[S] ratio and becomes independent of absolute concentration of ligand as the concentration of total ligand increases.

b. However, reaction rate *increases* with an increase in the [A]/[S] ratio at low concentrations of total ligand but *decreases* with an increase in the [A]/[S] ratio at high concentrations of total ligand. Why does this reversal occur?

c. Examine the table above and confirm that the reaction rate becomes a function of the [A]/[S] ratio (and independent of actual ligand concentrations) at a much lower total concentration (relative to $S_{0.5}$) for a cooperative enzyme than for a noncooperative enzyme. Why?

Substrate Analogs

3. Use the model described in Figure 4.34 to complete and graph the table below. Assume that $S_{0.5}$ has a value of 0.1, and that [S] is 0.04.

[Analog]	Without Analog	Reaction Rate With Analog	% Effect of Analog
0			
0.005			
0.01			
0.03			
0.05			
0.1			
0.3			
0.5			
1			
3			
5			
10			

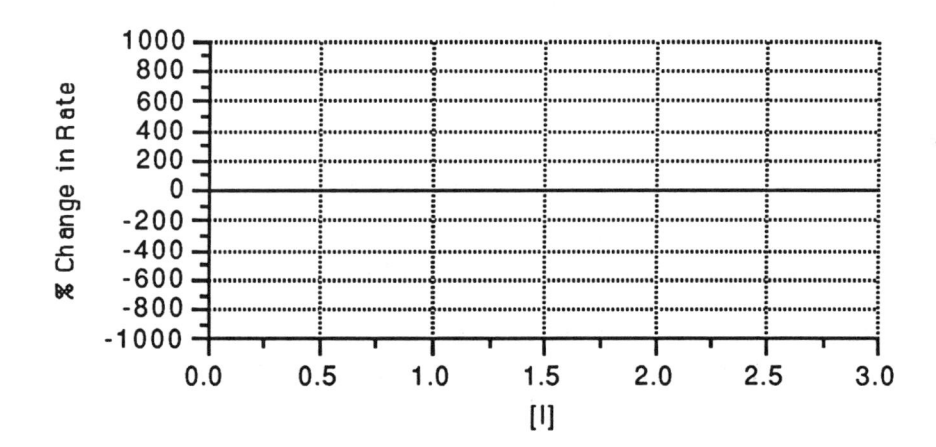

a. For cooperative enzymes, is it appropriate to regard an analog that competes with substrate for the catalytic site as a competitive inhibitor?

b. Can you explain in language that would be clear to a bright high school student why reaction rate first increases and then decreases as the concentration of competing analog increases?

163

Chapter 4   Experiment 4.13*

Problems

1. Set the concentration of substrate at different values and observe reaction rate as a function of analog concentration.

   Compare the absolute reaction velocities at different concentrations of competing analog when increasing values of [S], beginning at 0.01. What happens to: (1) the absolute reaction rate, and (2) the fold-increase in reaction rate induced by analog?

2. You might find it interesting to build a model of a cooperative enzymic reaction in the presence of a competitive substrate analog in which the degree of cooperativity is assignable, or in which different values can be assigned to $S_{0.5}$ and $A_{0.5}$, or in which which both of these changes can be made.

Chapter 5

Metabolism

Metabolism involves issues of stoichiometry, thermodynamics, and kinetics all working together. In this chapter you will build electronic spreadsheet models that link many of the issues developed in previous chapters into computer simulations. These simulations may help you develop a feeling for some of the interactions of metabolism that are difficult to acquire from textbooks, lectures, or unaided thought.

Electronic spreadsheets are ideal for such studies because

1. The structures of electronic spreadsheet models remain exposed for easy understanding.

2. Electronic spreadsheet models are easily modified to permit exploration of new combinations of parameters.

3. Electronic spreadsheets encourage active, undirected explorations in multiple directions that can reveal surprising interactions.

This chapter contains a range of models — from the very simple to the relatively advanced — that illustrates several important aspects of metabolism. However, if you have enjoyed the previous exercises in this text, now is the time to begin (if you have not already done so) to take other problems that have arisen in your studies and to build models that can assist you in their solution.

Perhaps this chapter will serve as your launch pad to more searching explorations.

Chapter 5   Experiment 5.1

Experiment 5.1

Simple Equilibrium: Review and New Considerations

In Experiment 3.1 you explored the behavior of a simple reaction with two precursors and two products (Figure 5.1).

In this experiment you will perform an additional exercise with the model you built in Experiment 3.1 (Figure 3.1), and you will build and explore a model designed to simulate a reaction with a single reactant and two products (Figure 5.2). You may be surprised at the differences in behavior exhibited by these two different situations.

In reactions where the number of precursors equals the number of products, the equilibrium constant K_{eq} is a pure number without dimensions (Figure 5.3a). However, in cases where the number of precursors does not equal of the number of products, K_{eq} may be regarded as having the dimensions of concentration or inverse concentration (Figure 5.3b).*

Figure 5.1

$$A + B \underset{}{\overset{K_{eq}}{\rightleftharpoons}} C + D$$

Figure 5.2

$$A \underset{}{\overset{K_{eq}}{\rightleftharpoons}} B + C$$

---

* By international convention, equilibrium constants are protected from dimensions by arbitrarily considering activities to be dimensionless. It is sometimes convenient to ignore this convention for reasons that will be clear to you when you complete this experiment.

Simple Equilibrium

Figure 5.3

(a)

$$A + B \underset{}{\overset{K_{eq}}{\rightleftharpoons}} C + D$$

$$K_{eq} = \frac{[C]_{eq}[D]_{eq}}{[A]_{eq}[B]_{eq}} = \frac{\text{Molar} \times \text{Molar}}{\text{Molar} \times \text{Molar}} = \text{(Dimensionless)}$$

(b)

$$A \underset{}{\overset{K_{eq}}{\rightleftharpoons}} B + C$$

$$K_{eq} = \frac{[B]_{eq}[C]_{eq}}{[A]_{eq}} = \frac{\text{Molar} \times \text{Molar}}{\text{Molar}} = \text{Molar}$$

The significance of equilibrium constants with the dimension of molarity, for example, becomes apparent when studying the aldolase-catalyzed cleavage of fructose bisphosphate (Figure 5.4).

It is often said that the equilibrium constant for this reaction is highly unfavorable ($K_{eq} = 10^{-4}$). However, this equilibrium constant has the dimensions of concentration, and the fraction of fructose bisphosphate

Figure 5.4

$$\text{FBP} \underset{}{\overset{K_{eq}}{\rightleftharpoons}} \text{DHAP} + \text{3-PGAld}$$

$$K_{eq} = \frac{[\text{DHAP}]_{eq}[\text{3-PGAld}]_{eq}}{[\text{FBP}]_{eq}} = \frac{\text{Molar} \times \text{Molar}}{\text{Molar}} = 10^{-4}M$$

where FBP = Fructose-1,6-*bis*PO$_4$
DHAP = Dihydroxyacetone-PO$_4$
3-PGAld = 3-PO$_4$-glyceraldehyde

Chapter 5   Experiment 5.1

cleaved at equilibrium depends not only on K_{eq} but also on the concentration of fructose diphosphate.

Directions

1. Build and verify the models described in Figure 3.1 (Experiment 3.1) and Figure 5.5.

2. The model described in Figure 5.5 is designed to simulate the type of reactions displayed in Figures 5.2 and 5.4. The design of the model is developed by a process identical to that already described in Experiment 3.1.

deb168 first example constructed on Migent Ability Spreadsheet (NOT LOTUS).

Figure 5.5

Formulas

	A	B	C	D	E
1	K =	0.0001			
2	[A]o =	0.1		[A]eq =	=B2-B10
3	[B]o =	0.001		[B]eq =	=B3+B10
4	[C]o =	0.002		[C]eq =	=B4+B10
5					
6	a =	1		[B]eq/[A]eq =	=E3/E2
7	b =	=B1+B3+B4			
8	c =	=(B3*B4)-(B1*B2)			
9					
10	x =	=(-B7+SQRT((B7*B7)-(4*B6*B8)))/(2*B6)			
11	x/[A]o =	=B10/B2			

(Continued.)

Figure 5.5 (Contd.)

Values

	A	B	C	D	E
1	K =	0.0001			
2	[A]o =	0.1		[A]eq =	0.0983
3	[B]o =	0.001		[B]eq =	0.0027
4	[C]o =	0.002		[C]eq =	0.0037
5					
6	a =	1		[B]eq/[A]eq =	0.0272
7	b =	0.0031			
8	c =	0.0000			
9					
10	x =	0.0017			
11	x/[A]o =	0.0168			

Questions

1. In a reaction of the sort described in Figure 5.1, K_{eq} is a dimensionless number, and the ratio of products to reactants is not affected by the total concentration of reactants plus products.

    If the initial concentrations of the products C and D are zero, then the fraction of A that is converted to C at equilibrium will not be changed if the initial concentrations of A and B are changed by the same factor.

    Use the model described in Figure 3.1 (Experiment 3.1) to complete the table below and confirm this fact.

$[A]_o$	$[B]_o$	$[C]_o$	$[D]_o$	$[C]_{eq}/[A]_o$
1.00	10.00	0.00	0.00	_____
5.00	50.00	0.00	0.00	_____
10.00	100.00	0.00	0.00	_____
50.00	500.00	0.00	0.00	_____

2. In a reaction of the sort described in Figure 5.2, K_{eq} has the dimensions of molarity, and the concentration of the reactant plays a significant role in the extent of the reaction.

Chapter 5   Experiment 5.1

a. Derive the mathematics used to build the model described in Figure 5.5. The plan of attack is identical to that already developed for Experiment 3.1.

b. Explore the aldolase-catalyzed cleavage of fructose bisphosphate (Figure 5.4) by completing the table below.

Although K_{eq} for this reaction is $10^{-4}M$ (0.1 mM), does the reaction appear unfavorable when the initial concentration of FBP is $10^{-4}M$?

At what initial concentration of FBP does the equilibrium ratio of [3-PGAld]/[FBP] equal one?   .0002 m

[FBP]$_o$	[3-PGAld]$_{eq}$	[3-PGAld]$_{eq}$/[FBP]$_{eq}$
1.0M	.01	.0101
0.1M	.003	.0321
0.01M	.003	.105
0.001M	.003	.372
0.0001M	.001	1.681

Problems

The two products of the aldolase reaction described in Question 2 are interconverted by the action of the enzyme triose phosphate isomerase.

$$\text{DHAP} \xrightleftharpoons{K_{eq}} \text{3-PGAld}$$

$$K_{eq} = \frac{[\text{3-PGAld}]_{eq}}{[\text{DHAP}]_{eq}} = 1/23$$

This reaction affects the ratios of the three compounds involved in the aldolase reaction. The model described in Figure 5.6 shows the effect of triose phosphate isomerase on the relative concentrations of the three sugar phosphates.

You will notice that the calculations in the two models (Figures 5.5 and 5.6) differ. Both approaches are useful, depending on the use to which the calculation is put. Figure 5.5 uses the standard chemical approach in which the system contains a fixed amount of material so that some concentrations must decrease as others increase. Most metabolic systems are near a steady state (i.e., concentrations are constant although reaction is occurring), and the simpler approach in Figure 5.6 is generally more appropriate.

Does the effect of the isomerase on the [3-PGAld]/[FBP] ratio change with the concentration of FBP? Why or why not?

Phosphoglyceraldehyde is the substrate for the later stages of glycolysis; under ordinary circumstances the function of the isomerase is to convert DHAP to 3-PGAld so that it can be further metabolized. Its small negative effect on the [3-PGAld]/[FBP] ratio is therefore unimportant in view of its essential role in the metabolism of half the carbon atoms of glucose and other hexoses.

Figure 5.6

Formulas

	A	B	C	D
1	[FBP] =	1000		mM
2				
3	Triose-PO4-Isomerase:	Absent	Present	
4	[3-PGAld] =	=SQRT(0.1*B1)	=SQRT(0.1*B1/23)	mM
5	[DHAP] =	=B4	=23*C4	mM
6	[3-PGAld]/[FBP] =	=B4/B1	=C4/B1	mM

Values

	A	B	C	D
1	[FBP] =	1000		mM
2				
3	Triose-PO4-Isomerase:	Absent	Present	
4	[3-PGAld] =	10.0000	2.0851	mM
5	[DHAP] =	10.0000	47.9583	mM
6	[3-PGAld]/[FBP] =	0.0100	0.0021	mM

Experiment 5.2

Multiple Sequential Equilibria

Consider a sequence of reactants in equilibrium with each other:

$$A \rightleftharpoons B \rightleftharpoons C \rightleftharpoons D \rightleftharpoons P$$

Most reaction sequences contain more than one reactant at each step, but this simplified model still illustrates the principles that govern multiple sequential equilibria.

In a sequence of reactions, the overall change in Gibbs standard free energy determines the equilibrium ratio of the concentration of the end product to the concentration of the starting compound, $[P]_{eq}/[A]_{eq}$. You might think, therefore, that only the overall standard free-energy change (between A and P) is important and that the free energies of formation of the intermediates play no role in the equilibrium configuration.

This is not correct. The ratio $[P]_{eq}/[A]_{eq}$ is indeed determined only by the difference between the standard free energies of formation of A and P (assuming that there are no reactants or products other than those shown), but the actual concentration of P at equilibrium, relative to the total concentration $[A] + [B] + [C] + [D] + [P]$, varies sharply with changes in the free energies of formation of the intermediates.

In this experiment you will see that intermediates with free energies of formation that are only slightly lower than the end product can accumulate to surprisingly high concentrations.

Such considerations are important in the study of metabolism because the accumulation of intermediates can have undesirable effects (side reactions, decreased solvent capacity of the cell, etc.).

Consider the change in free energy that exists between A and B:

$$\Delta G° = G°(B) - G°(A) = -RT \ln \frac{[B]_{eq}}{[A]_{eq}} = -RT \ln(K_{eq}) \quad (5.1)$$

Where $\Delta G°$ = standard free energy change of the reaction; difference between the standard free energies of formation of B and A (kcal/mol)
$G°(B)$ = standard free energy of formation of B (kcal/mol)
$G°(A)$ = standard free energy of formation of A (kcal/mol)
R = the gas constant (1.98 cal/°K mol)
T = temperature (°K)
$[B]_{eq}$ = the concentration of reactant B at equilibrium (M)
$[A]_{eq}$ = the concentration of reactant A at equilibrium (M)

Solving for $[B]_{eq}/[A]_{eq}$ yields:

$$\frac{[B]_{eq}}{[A]_{eq}} = e^{\left(\frac{-\Delta G°}{RT}\right)} = e^{\left(\frac{G°(A)-G°(B)}{RT}\right)} \quad (5.2)$$

The equations for $[C]_{eq}/[B]_{eq}$, $[D]_{eq}/[C]_{eq}$, and $[P]_{eq}/[D]_{eq}$ are similarly constructed:

$$\frac{[C]_{eq}}{[B]_{eq}} = e^{\left(\frac{-\Delta G°}{RT}\right)} = e^{\left(\frac{G°(B)-G°(C)}{RT}\right)} \quad (5.3)$$

$$\frac{[D]_{eq}}{[C]_{eq}} = e^{\left(\frac{-\Delta G°}{RT}\right)} = e^{\left(\frac{G°(C)-G°(D)}{RT}\right)} \quad (5.4)$$

$$\frac{[P]_{eq}}{[D]_{eq}} = e^{\left(\frac{-\Delta G°}{RT}\right)} = e^{\left(\frac{G°(D)-G°(P)}{RT}\right)} \quad (5.5)$$

$[A]_{eq}$, expressed as a percentage of the total concentration of all reactants, can be calculated from the ratios obtained in Equations 5.2 through 5.5 by using the partition equation:

$$100 = [A]_{eq} + [B]_{eq} + [C]_{eq} + [D]_{eq} + [P]_{eq} \quad (5.6)$$

or

$$100 = [A]_{eq}\left[1 + \left(\frac{[B]_{eq}}{[A]_{eq}}\right) + \left(\frac{[C]_{eq}}{[B]_{eq}}\right)\left(\frac{[B]_{eq}}{[A]_{eq}}\right)\right.$$
$$+ \left(\frac{[D]_{eq}}{[C]_{eq}}\right)\left(\frac{[C]_{eq}}{[B]_{eq}}\right)\left(\frac{[B]_{eq}}{[A]_{eq}}\right) \quad (5.7)$$
$$\left. + \left(\frac{[P]_{eq}}{[D]_{eq}}\right)\left(\frac{[D]_{eq}}{[C]_{eq}}\right)\left(\frac{[C]_{eq}}{[B]_{eq}}\right)\left(\frac{[B]_{eq}}{[A]_{eq}}\right)\right]$$

Equation 5.7 is solved for $[A]_{eq}$, and the concentrations of the other compounds are calculated from $[A]_{eq}$ and the previously computed concentration ratios (Equations 5.2 through 5.5):

$[B]_{eq} = [A]_{eq}([B]_{eq}/[A]_{eq})$ (5.8)
$[C]_{eq} = [B]_{eq}([C]_{eq}/[B]_{eq})$ (5.9)
$[D]_{eq} = [C]_{eq}([D]_{eq}/[C]_{eq})$ (5.10)
$[P]_{eq} = [D]_{eq}([P]_{eq}/[D]_{eq})$ (5.11)

Thus, given the free energies of formation of A, B, C, D, and P, it is possible to compute the relative concentrations of these reactants at equilibrium.

Directions

1. Build and verify the model described in Figure 5.7. Be sure to set the recalculation order to columnwise.

2. Refer back to Experiment 3.3 (Figure 3.4) and note the similarities between the two models. Arginine ionization is specific example of multiple sequential equilibria.

Figure 5.7

Formulas

	A	B	C	D	E
1	G°(A) =	0		R =	0.001987
2	G°(B) =	-1		T =	300
3	G°(C) =	-2			
4	G°(D) =	-2			
5	G°(P) =	-4			
6					
7	[B]eq/[A]eq =	=EXP((B1-B2)/(E1*E2))			
8	[C]eq/[B]eq =	=EXP((B2-B3)/(E1*E2))			
9	[D]eq/[C]eq =	=EXP((B3-B4)/(E1*E2))			
10	[P]eq/[D]eq =	=EXP((B4-B5)/(E1*E2))			
11					
12	100/A =	=(1+B7+(B8*B7)+(B9*B8*B7)+(B10*B9*B8*B7))			
13					
14	[A]eq =	=100/B12		[B]eq/[A]eq =	=B15/B14
15	[B]eq =	=B14*B7		[C]eq/[A]eq =	=B16/B14
16	[C]eq =	=B15*B8		[D]eq/[A]eq =	=B17/B14
17	[D]eq =	=B16*B9		[P]eq/[A]eq =	=B18/B14
18	[P]eq =	=B17*B10			

Values

	A	B	C	D	E
1	G°(A) =	0		R =	0.001987
2	G°(B) =	-1		T =	300
3	G°(C) =	-2			
4	G°(D) =	-2			
5	G°(P) =	-4			
6					
7	[B]eq/[A]eq =	5.3525			
8	[C]eq/[B]eq =	5.3525			
9	[D]eq/[C]eq =	1			
10	[P]eq/[D]eq =	28.6497			
11					
12	100/A =	884.4552			
13					
14	[A]eq =	0.1131		[B]eq/[A]eq =	5.3525
15	[B]eq =	0.6052		[C]eq/[A]eq =	28.6497
16	[C]eq =	3.2392		[D]eq/[A]eq =	28.6497
17	[D]eq =	3.2392		[P]eq/[A]eq =	820.8033
18	[P]eq =	92.8033			

Recalculation Order = COLUMNWISE

Chapter 5   Experiment 5.2

Questions

In Figure 5.7, $G°(P)$ is 4 kcal/mol less than $G°(A)$, $[P]_{eq}/[A]_{eq}$ equals 821, and $[P]_{eq}$ constitutes 92.8% of all of the reactants combined. Interrogate the model described in Figure 5.7 by completing the table below.

$G°(C)$	$[A]_{eq}$	$[C]_{eq}$	$[P]_{eq}$	$[P]_{eq}/[A]_{eq}$
-2				
-3				
-4				
-5				
-6				
-7				

a. As the free energy of formation of C decreases from -2 kcal/mol to -7 kcal/mol what happens to the equilibrium concentration of P?

b. Does $[P]_{eq}/[A]_{eq}$ change?

c. What happens to the mass that used to reside in P?

Your results show why metabolic sequences have evolved to avoid intermediates with low values of standard free energy of formation. (Because there are usually other reactants and products in addition to the linear intermediates, a more general statement is that reactions with large positive values of $\Delta G°$ are avoided.)

Experiment 5.3

Kinetic Analysis of Sequential Reactions

Consider the forward and reverse reactions of two reactants in equilibrium with each other:

$$A \underset{k_{-1}}{\overset{k_1}{\rightleftharpoons}} B$$

If the reactions fit first-order kinetics, the initial reaction rates v_1 (the rate at which A converts to B) and v_{-1} (the rate at which B converts to A) are given by the rate laws

$$v_1 = k_1[A] \tag{5.12}$$
$$v_{-1} = k_{-1}[B] \tag{5.13}$$

The net conversion (or flux, F) from A to B is given by:

$$F = k_1[A] - k_{-1}[B] \tag{5.14}$$

For a system in equilibrium, the forward and reverse reaction rates are equal, and so $F = 0$:

$$F = k_1[A] - k_{-1}[B] = 0 \tag{5.15}$$

or

$$[B] = \frac{k_1[A]}{k_{-1}} \tag{5.16}$$

The above equilibrium can, of course, be expanded over as many reactants as desired:

$$A \underset{k_{-1}}{\overset{k_1}{\rightleftharpoons}} B \underset{k_{-2}}{\overset{k_2}{\rightleftharpoons}} C \underset{k_{-3}}{\overset{k_3}{\rightleftharpoons}} D \underset{k_{-4}}{\overset{k_4}{\rightleftharpoons}} P$$

When the sequence is in equilibrium, each component reaction must also be in equilibrium; thus Equations 5.15 and 5.16 apply to each of the reactions in the sequence of reactions:

$$[B] = \frac{k_1[A]}{k_{-1}} \quad [C] = \frac{k_2[B]}{k_{-2}} \quad [D] = \frac{k_3[C]}{k_{-3}} \quad [P] = \frac{k_4[D]}{k_{-4}} \quad (5.17 - 5.20)$$

When the sequence exhibits a *steady state* net flow or flux in the forward direction, flux must have the same value for all of the component reactions:

$$F = k_1[A] - k_{-1}[B] = k_2[B] - k_{-2}[C] = k_3[C] - k_{-3}[D] = k_4[D] - k_{-4}[P] \quad (5.21)$$

The rate expression for each reaction can be solved for the concentration of its product:

$$[B] = \frac{k_1[A] - F}{k_{-1}} \quad [C] = \frac{k_2[B] - F}{k_{-2}}$$
$$[D] = \frac{k_3[C] - F}{k_{-3}} \quad [P] = \frac{k_4[D] - F}{k_{-4}} \quad (5.22 - 5.25)$$

These relationships may be used to construct a model of a sequence of reactions in which the flux from precursor to product varies with the rate at which the product is removed.

In effect, this model calculates the concentration of the end product P and each of the intermediates when the flux is at steady state (i.e., the concentrations of each of the intermediates is constant). For true steady state to exist, it is necessary that A be supplied and P be removed at rates such that their concentrations remain constant. This situation is common in metabolism but is relatively rare in nonbiological situations.

Kinetic Analysis of Sequential Reactions

Remember that metabolic reactions are catalyzed by enzymes so that the simple relationships of Equations 5.12 and 5.13 do not apply. In this experiment you will explore the nonenzymic case; in the following experiment you will consider enzyme-catalyzed reactions.

Directions

Build and verify the model described in Figure 5.8.

Figure 5.8

Formulas

	A	B	C	D	E	F	G
1	F =	0.2		k1 =	1.0	k1[A] =	=E1*B2
2	[A] =	10		k-1 =	0.1	k-1[B] =	=E2*B4
3				k2 =	0.2	k2[B] =	=E3*B4
4	[B] =	=(E1*B2−B1)/E2		k-2 =	0.2	k-2[C] =	=E4*B5
5	[C] =	=(E3*B4−B1)/E4		k3 =	0.1	k3[C] =	=E5*B5
6	[D] =	=(E5*B5−B1)/E6		k-3 =	1.0	k-3[D] =	=E6*B6
7	[P] =	=(E7*B6−B1)/E8		k4 =	0.5	k4[D] =	=E7*B6
8				k-4 =	0.1	k-4[P] =	=E8*B7

Values

	A	B	C	D	E	F	G
1	F =	0.2		k1 =	1.0	k1[A] =	10.0
2	[A] =	10		k-1 =	0.1	k-2[B] =	9.8
3				k2 =	0.2	k2[B] =	19.6
4	[B] =	98		k-2 =	0.2	k-2[C] =	19.4
5	[C] =	97		k3 =	0.1	k3[C] =	9.7
6	[D] =	9.5		k-3 =	1.0	k-3[D] =	9.5
7	[P] =	45.5		k4 =	0.5	k4[D] =	4.8
8				k-4 =	0.1	k-4[P] =	4.6

Recalculation Order = COLUMNWISE

Chapter 5   Experiment 5.3

Questions

1. Use the model described in Figure 5.8 to complete and graph the table below. Assume that [A] = 10.

F	[B]	[C]	[D]	[P]
0.2	___	___	___	___
0.4	___	___	___	___
0.6	___	___	___	___
0.8	___	___	___	___
1.0	___	___	___	___
1.2	___	___	___	___
1.4	___	___	___	___
1.6	___	___	___	___
1.8	___	___	___	___
2.0	___	___	___	___

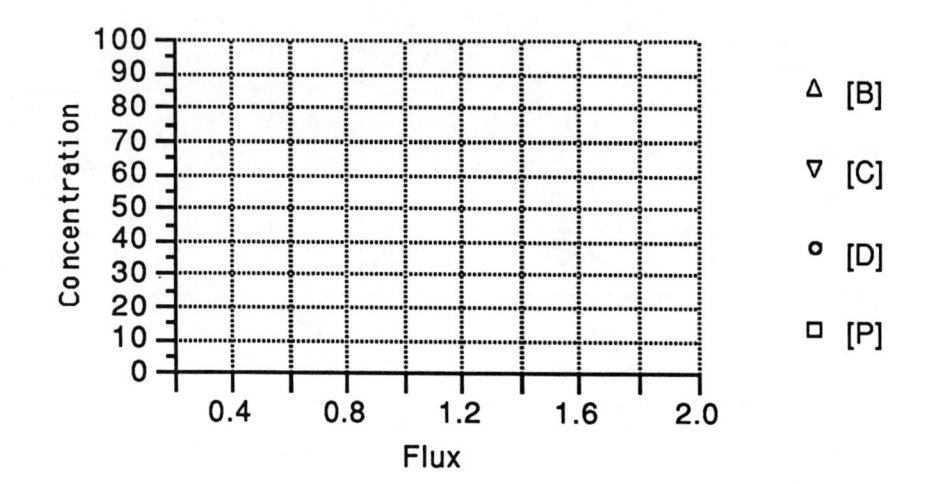

a. As the flux F increases in value, do the concentrations of P and all intermediates increase or decrease?

b. Rate laws such as those in Equations 5.12 and 5.13 state that reaction rates *increase* as precursor concentrations *increase*. How do you justify the fact that the rate of conversion of C to D increases, for example, as [C] decreases?

Kinetic Analysis of Sequential Reactions

2. The answer to Question 1b may be approached by exploring cells G1 to G8 of the model described in Figure 5.8. Complete and graph the table below.

F	k_2[B]	k_{-2}[C]	k_4[D]	k_{-4}[P]
0.2	___	___	___	___
0.4	___	___	___	___
0.6	___	___	___	___
0.8	___	___	___	___
1.0	___	___	___	___
1.2	___	___	___	___
1.4	___	___	___	___
1.6	___	___	___	___
1.8	___	___	___	___
2.0	___	___	___	___

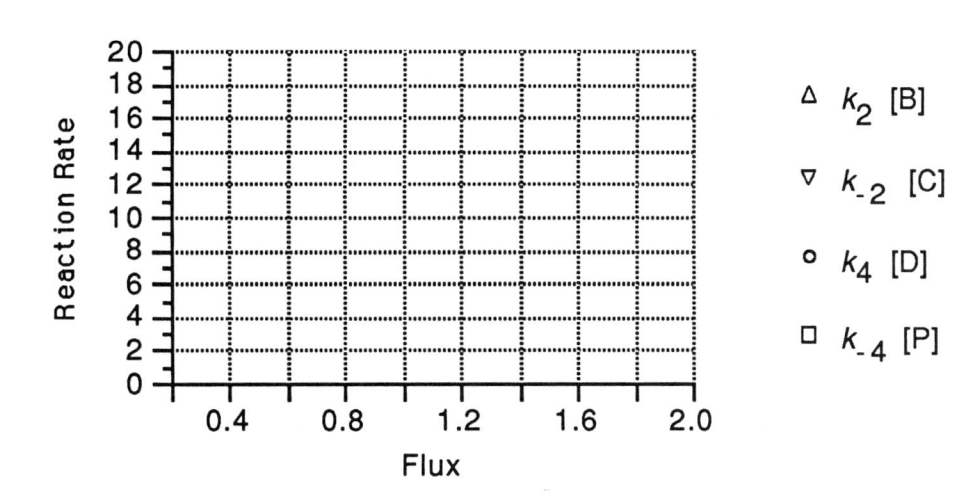

△ k_2 [B]

▽ k_{-2} [C]

○ k_4 [D]

□ k_{-4} [P]

a. You can obtain the concentrations of P and the intermediates at equilibrium by setting flux to zero. How do the concentrations change as the net flux is increased? As flux increases, how does the forward reaction rate B › C change? How does the reverse reaction C › B change?

b. Does the answer to part a above clarify the situation described in Question 1?

Chapter 5   Experiment 5.3

3. Net flux increases as [P] decreases (catalysts and other conditions remaining constant). In this model F is considered to be the independent variable and [P] the dependent variable. For most non-metabolic cases the relationship is reversed and flux depends on the concentrations of A and P.

4. As discussed above in Question 3, the model described in Figure 5.8 treats flux as the independent variable and [P] as the dependent variable. Now consider flux as a function of the rate constants and of [A] and [P].

Using Equation 5.25 it is possible to solve for [D] as a function of [P] and F. This value can be substituted into Equation 5.24 to obtain [C] as a function of [P] and F. Continuing this process through Equation 5.22 yields an equation that expresses [A] as a function of [P] and F. This equation can be rearranged such that F may be calculated from [A], [P], and the rate constants:

$$F = \frac{k_1 k_2 k_3 k_4 [A] - k_{-1} k_{-2} k_{-3} k_{-4} [P]}{k_2 k_3 k_4 + k_{-1} k_3 k_4 + k_{-1} k_{-2} k_4 + k_{-1} k_{-2} k_{-3}}$$

Build and verify the model below and then answer the following questions.

dyn182.wks.

Formulas

	A	B	C	D	E
1	[A] =	10		k1 =	1
2	[P] =	5		k-1 =	0.1
3				k2 =	0.2
4				k-2 =	0.2
5	Flux =	=((E1*E3*E5*E7*B1)-(E2*E4*E6*E8*B2))/ ((E3*E5*E7)+(E2*E5*E7)+(E2*E4*E7)+(E2*E4*E6))		k3 =	0.1
6	[B] =	=((E1*B1)-B5)/E2		k-3 =	1
7	[C] =	=((E3*B6)-B5)/E4		k4 =	0.5
8	[D] =	=((E5*B7)-B5)/E6		k-4 =	0.1

Values

	A	B	C	D	E
1	[A] =	10		k1 =	1
2	[P] =	5		k-1 =	0.1
3				k2 =	0.2
4				k-2 =	0.2
5	Flux =	2		k3 =	0.1
6	[B] =	80		k-3 =	1
7	[C] =	70		k4 =	0.5
8	[D] =	5		k-4 =	0.1

a. If [A] is held constant, how is flux affected by variation in the concentration of product [P]?

At high values of [P], the value of F is negative. What does this mean?

What is the significance of [P] at which F equals zero?

5. Consider further the model built for Question 4. What would limit F if all reactions had extremely large values of K_{eq}? (**Hint:** Set all reverse rate constants to zero. You may be surprised to find that F under these circumstances depends on only one of the four rate constants. How is that possible? Why would k_4, for example, not be limiting if its value were much smaller than those of the other rate constants?

6. What would limit F if P were swept away as rapidly as it was produced, so that its effective concentration was held at zero?

7. If [P] and [A] are held constant, how is the flux affected by doubling the forward and reverse rate constants between A and B (e.g., by an increase in the concentration of a catalyst)?

What if the rate constants between C and D were doubled? Why the difference?

What generalization is illustrated, that is, why is the flux more sensitive to changes in the rate constants for some reactions than to those for others?

8. Note how the concentrations of the reaction intermediates are affected by changes in the amounts of individual catalysts. (Change the values of the rate constants for different steps by the same factor in both directions and observe the results.)

Experiment 5.4

Substrate Concentration as a Function of V_{max}, K_m, and Reaction Velocity

In the laboratory, kinetic experiments with enzymes typically involve measurements of reaction rates at selected substrate concentrations. It is therefore natural to think of substrate concentration as the independent variable and reaction rate as the dependent variable. In fact, the opposite is true for most enzymes in the intact cell.

The flux through a typical reaction sequence or metabolic pathway is controlled by the rate of the first reaction in the sequence. This reaction rate is controlled by the regulatory enzyme at the beginning of the sequence. The reaction rates of intermediate steps of the reaction sequence are governed by the flux through the sequence as a whole, and the concentrations of the intermediates must adjust accordingly (Figure 5.9).

Figure 5.9

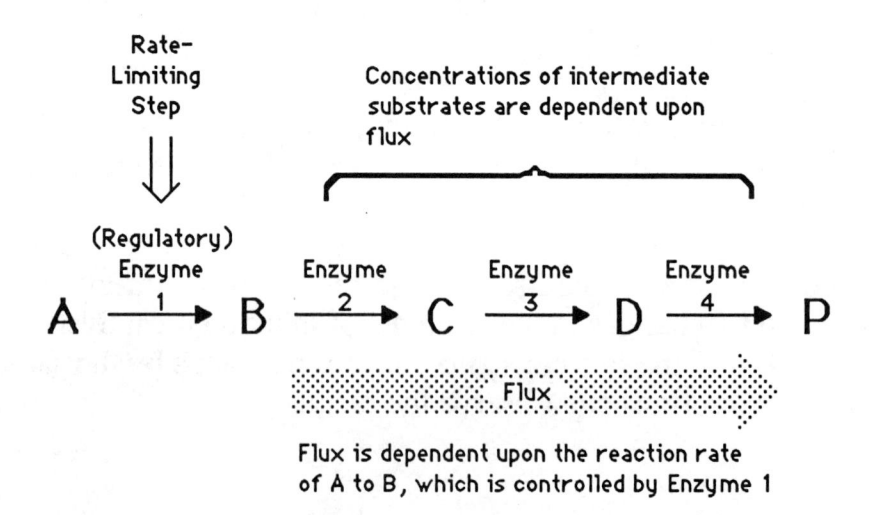

Rate-Limiting Step ⇓ (Regulatory) Enzyme

Concentrations of intermediate substrates are dependent upon flux

$$A \xrightarrow{1} B \xrightarrow{2} C \xrightarrow{3} D \xrightarrow{4} P$$

Flux

Flux is dependent upon the reaction rate of A to B, which is controlled by Enzyme 1

Substrate Concentration, Vmax, Km, and Reaction Velocity

Such behavior is readily modeled by simply rearranging the traditional form of the Michaelis equation (Equation 5.26) such that [S] becomes a function of K_m, v, and V_{max} (Equation 5.27).

$$\frac{v}{V_{max}} = \frac{[S]}{K_m + [S]} \tag{5.26}$$

$$[S] = \frac{K_m v}{V_{max} - v} \tag{5.27}$$

Directions

Build and verify the model described in Figure 5.10.

Figure 5.10

Formulas

	A	B
1	Flux =	100
2	Vmax =	200
3	Km =	1
4		
5	[S] =	=(B3*B1)/(B2-B1)

Values

	A	B
1	Flux =	100
2	Vmax =	200
3	Km =	1
4		
5	[S] =	1

Chapter 5   Experiment 5.4

Questions

1. Use the model described in Figure 5.10 to complete and graph the table below.

Flux	$V_{max} = 100; K_m = 1$ [S]	$V_{max} = 200; K_m = 1$ [S]
0	_____	_____
20	_____	_____
40	_____	_____
60	_____	_____
80	_____	_____
90	_____	_____
95	_____	_____
99	_____	_____
95	_____	_____
99	_____	_____
120	_____	_____
140	_____	_____
160	_____	_____
180	_____	_____
190	_____	_____
198	_____	_____

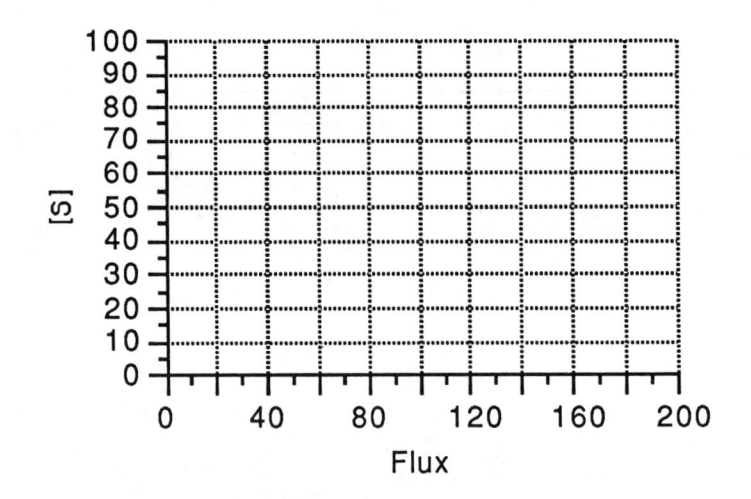

▲ $V_{max} = 100$

▼ $V_{max} = 200$

a. When $V_{max} = 100$, what happens to [S] as flux approaches 100? When $V_{max} = 200$?

This simulation is relevant to the concentrations of intermediates along a metabolic pathway. If the flux were to increase to levels where it exceeds the catalytic capacity of the enzyme, the concentration of the substrate of that enzyme would rise

Substrate Concentration, Vmax, Km, and Reaction Velocity

catastrophically without limit unless it was excreted by the cell or converted to a side product (which may well be toxic).

2. Use the model described in Figure 5.10 to complete and graph the table below.

V_{max}	Flux = 100; K_m = 1 [S]	Flux = 200; K_m = 1 [S]
1000	____	____
800	____	____
600	____	____
400	____	____
300	____	____
220	____	____
210	____	____
202	____	____
200	____	
150	____	
110	____	
105	____	
101	____	

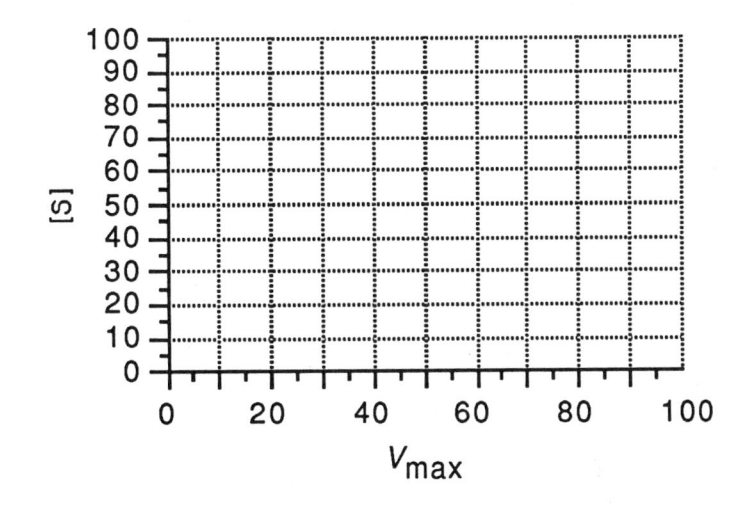

▲ Flux = 100

▼ Flux = 200

a. When flux = 100, what happens to [S] as the enzyme concentration decreases such that V_{max} approaches 100?

This would be the effect of a decreasing enzyme amount on the concentration of an intermediate in a reaction sequence where flux is constant. Do you see the importance of mechanisms that regulate enzyme levels (e.g., induction, repression)?

Chapter 5   Experiment 5.4

3. It is often assumed that reactions in intact cells proceed at rates close to the V_{max} of their enzymes. Your study of the previous two questions should already cause you to doubt this assumption.

   a. Use the model described in Figure 5.10 (modify it slightly if you wish) to determine the percent increase in [S] that occurs when a flux of 0.85 of V_{max} increases by 12% to 0.95 of V_{max}.   [S] = 19
      [S] = 5.7

   b. Now determine the percent increase in [S] that occurs when a flux of 0.30 of V_{max} increases by 12% to 0.34 of V_{max}.
      [S] = 0.42   [S] = 0.51

Problems

1. Consider a metabolic pathway in which flux is 85% of V_{max} for an enzyme that catalyzes the reaction of an intermediate in the pathway. What is the margin of safety for increases in the rate at which the initial metabolite enters the sequence if the concentration of the intermediate may not rise more than 50% without damaging the cell? What is the margin of safety for increases in flux if the intermediate may not increase by more than two-fold? five-fold?

   85 → 88.5%   85 → 92%   85 → 96.5%

2. Repeat Problem 1 while considering a metabolic pathway in which the flux is 35% of V_{max}.

   35 → 45% (0.5×S)   35 → 50 (2×S)   35 → 73% (5×S)

3. It is often noted that the concentrations of many metabolites in living cells are somewhat below the K_m values of the enzymes that act upon them. What determines the physiologic steady-state levels of intermediates in metabolic pathways? (The steady-state levels of these intermediates must be distinguished from the concentrations of branchpoint metabolites and end products that are usually controlled by specific and complex regulatory interactions.)

4. In agreement with the first sentence of Problem 3, the activities of most enzymes seem to be held at levels that result in normal fluxes of between 10% and 35% of V_{max}. By now you should realize that selection pressures in the course of evolution make such values reasonable. What is the concentration of substrate as a function of K_m when flux is 0.30 of V_{max}?

   ~0.42

Substrate Concentration, Vmax, Km, and Reaction Velocity

Determine the boundaries on the ratio of [S] to K_m (at "normal" values of flux) if the levels of enzymes are sufficient to allow the flux to increase three-fold above "normal" values without increasing the concentrations of the intermediates by more than a factor of eight.

Determine the boundaries on the ratio of [S] to K_m if the levels of enzymes are sufficient to allow the flux to increase four-fold without increasing [S] by more than a factor of ten?

5. How would you answer someone who argues that it is wasteful for a cell to make more enzyme than just enough to permit the reaction rates required by "normal" conditions?

Conclusion

In this experiment you have seen that it is important to the cell that enzyme levels be sufficiently high so that normal reaction velocities do not exceed about 30% of V_{max}. The amount of enzyme in a cell is a factor that interacts with other properties of enzymes, which will be considered elsewhere in this book. The systems that control levels of enzymes are outside the scope of this book.

Experiment 5.5

Competition between Enzymes at a Branch Point

Metabolism is regulated primarily by partitioning metabolites at branch points where, for example, a biosynthetic pathway branches from a major catabolic pathway. Two enzymes, one catalyzing the next reaction in the catabolic sequence and the other catalyzing the first step of the synthetic sequence, compete for the branch point metabolite that acts as substrate for both enzymes (Figure 5.11).

The rate at which the branch point metabolite (e.g., B in Figure 5.11) is produced is usually regulated by enzymes that catalyze earlier steps in the main pathway. The fraction of this flux that flows into the branching pathway is regulated by changes in the properties of the competing enzymes (e.g., B-ase I and B-ase II, Figure 5.11).

In end product inhibition, for example, the concentration of the product of the branch pathway controls the affinity of substrate for the first enzyme in the branch pathway (e.g., B-ase II, Figure 5.11). An example of

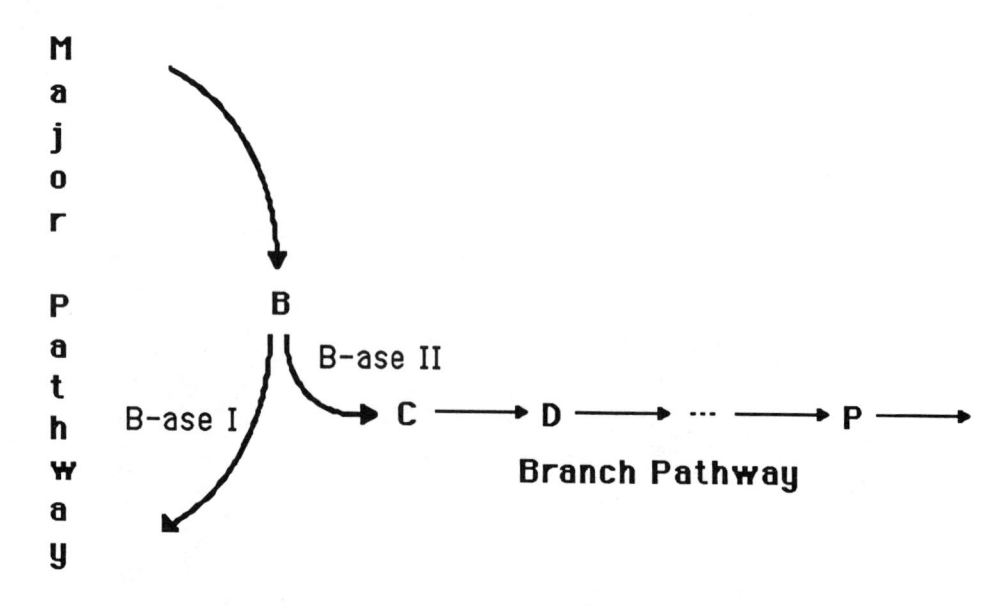

Figure 5.11

Major Pathway: B → (B-ase I) ↓

B → (B-ase II) → C → D → ··· → P →

Branch Pathway

Competition Between Enzymes at a Branch Point

end product inhibition is modeled in Experiment 5.6. In this experiment you will control partitioning by simply assigning values of $S_{0.5}$ for B-ase II.

If v_1 and v_2 are the reaction rates of B-ase I and B-ase II, respectively, then total flux F in the major pathway before the branch point will be given by

$$v_1 + v_2 = F \qquad (5.28)$$

If both B-ase I and B-ase II contain four catalytic sites and exhibit essentially infinite cooperativity, then v_1 and v_2 can be calculated as already noted in Experiment 4.7 as Equation 4.34. Equation 5.28 can then be expanded to yield

$$\frac{V_{max(1)}[S]^4}{[S_{0.5}]_1^4 + [S]^4} + \frac{V_{max(2)}[S]^4}{[S_{0.5}]_2^4 + [S]^4} = F \qquad (5.29)$$

Where the subscripts 1 and 2 in V_{max} and $[S]_{0.5}$ refer to B-ase I and B-ase II, respectively (Figure 5.11)

Equation 5.29 may be rearranged to yield Equation 5.30.

$$(V_{max(1)} + V_{max(2)} - F)[S]^8$$
$$+ (V_{max(1)}[S_{0.5}]_2^4 + V_{max(2)}[S_{0.5}]_1^4 - F([S_{0.5}]_1 + [S_{0.5}]_2)^4)[S]^4 \qquad (5.30)$$
$$- F([S_{0.5}]_1 [S_{0.5}]_2) = 0$$

Equation 5.30 can be solved as a quadratic equation in $[S]^4$ by the usual techniques (e.g., Experiment 3.1, Equation 3.7a), and the reaction velocities for B-ase I and B-ase II calculated from the two terms in Equation 5.29.

Directions

1. Build and verify the model described in Figure 5.12.

Chapter 5   Experiment 5.5

2. The model computes reaction velocities (and v/V_{max}) over a four-fold range of values for $S_{0.5}$. The change in v/V_{max} for a given change in $S_{0.5}$ is a useful measure of the system's sensitivity of control (cells F14, F15, and G15).

Figure 5.12

dyn192.wks

Formulas

	A	B	C	D	E	F	G
1	Vmax1 =	1000					
2	Vmax2 =	100					
3	Flux =	100					
4	Lowest S0.5 (enz 2) =	0.6					
5							
6							
7	S0.5 (enz2)	a	b	c	determ		
8	=B4	Formula Set I	Formula Set II	Formula Set III	Formula Set IV		
9	=2*A8					Change in v2 when S(0.5)	
10	=2*A9					increases by	a factor of:
11						2	4
12	S0.5 (enz2)	[S]	v1	v2	v2/Vmax2		
13	=A8	Formula Set V	Formula Set VI	Formula Set VII	Formula Set VIII		
14	=A9					=D13/D14	
15	=A10					=D14/D15	=D13/D15

Values

	A	B	C	D	E	F	G
1	Vmax1 =	1000					
2	Vmax2 =	100					
3	Flux =	100					
4	Lowest S0.5 (enz 2) =	0.6					
5							
6							
7	S0.5 (enz2)	a	b	c	determ		
8	0.6	1.0E3	1.2E2	-1.3E1	2.6E2		
9	1.2	1.0E3	1.9E3	-2.1E2	2.1E3	Change in v2 when S(0.5)	
10	2.4	1.0E3	3.0E4	-3.3E3	3.0E4	increases by	a factor of:
11						2	4
12	S0.5 (enz2)	[S]	v1	v2	v2/Vmax2		
13	0.6	0.51	65.06	34.94	0.35		
14	1.2	0.57	95.17	4.83	0.05	7.24	
15	2.4	0.58	99.67	0.33	0.00	14.52	105.06

(Continued.)

Competition Between Enzymes at a Branch Point

Figure 5.12 (Contd.)

Formula Set I
Prototype Cell is B8
=B1+B2-B3

Formula Set II
Prototype Cell is C8
=((A8^4)*B1)+(B2)-B3-((A8^4)*B3)

Formula Set III
Prototype Cell is D8
=-B3*(A8^4)

Formula Set IV
Prototype Cell is E8
=SQRT(C8*C8-(4*B8*D8))

Formula Set V
Prototype Cell is B13
=SQRT(SQRT((-C8+E8)/(2*B8)))

Formula Set VI
Prototype Cell is C13
=(B1*(B13^4))/((B13^4)+1)

Formula Set VII
Prototype Cell is D13
=(B2*(B13^4))/((B13^4)+(A13^4))

Formula Set VIII
Prototype Cell is E13
=D13/B2

Questions

1. Use the model described in Figure 5.12 to complete Table A at the end of this section. *Use the contents of cell G15 as your "sensitivity index,"* but observe the contents of cells F14 and F15 as well. In some cases the two two-fold indices are quite different, illustrating how sharply the sensitivity of control changes with [S] under some conditions.

   In Table A the rate of reaction and the sensitivity index vary in opposite directions. You will find that this is true for all other tables as well. No matter how the parameters are varied, a trade-off between reaction velocity and regulatory effect is inevitable. (It is possible, however, to optimize the system such that a sensitivity index of greater than 100 and a v/V_{max} of 0.3 to 0.4 may exist simultaneously.)

2. Complete Table B. If there is no change in the properties of the first enzyme in the branch pathway, how does a change in flux through

the main sequence affect the flow of metabolites into the branch pathway? How is sensitivity to control in the branch pathway affected?

3. In most cases the flux through the branch pathway will be small compared to that in the main pathway (e.g., Tables A and B).

   Complete Tables C and D to explore regulatory behavior at the branch point when flux through the branch pathway is equal to flux through the main pathway. Compare your results with those previously obtained in Tables A and B. You should see that when branch pathway reaction rates as a whole are increased, sensitivity to control is decreased.

4. In Question 3 you saw that control sensitivity decreases when the flux through the branch pathway increases with respect to the main pathway. Table E reflects a decrease in the flux through the branch pathway (with respect to Table A). Confirm that control sensitivity increases when the flux through the branch pathway decreases with respect to the main pathway. The value of v/V_{max}, of course, decreases as control sensitivity increases.

5. In the real world the relative amounts of B-ase I and B-ase II (Figure 5.11) tend to remain constant, and it is the activity of the the first enzyme in the branch pathway (B-ase II) that regulates the flow of metabolite through the branch pathway (Tables A and D). However, with computer simulation it is possible to directly test the conclusions of Questions 3 and 4 by creating an environment unlikely to occur in nature.

   Complete Table F and ponder the results. Sensitivity of the branch pathway to control by variations in $S_{0.5}$ of B-ase II is most effective under just those conditions that are most likely to occur in the living cell (i.e., the flow through the branch pathway is small when compared with that in the main pathway). Recall that variations in $S_{0.5}$ are usually caused by the binding of modifier to the enzyme (Experiments 4.7, 4.8, 4.10, and 4.12). The modifier molecule is often the end product of the branch pathway itself (Experiment 5.6).

Competition Between Enzymes at a Branch Point

TABLE A

$V_{max(1)}$	$V_{max(2)}$	F	Lowest $S_{0.5}$	v_2	$v_2/V_{max(2)}$	Sens. Index
1000	100	100	0.1			
1000	100	100	0.2			
1000	100	100	0.3			
1000	100	100	0.5			
1000	100	100	0.6			
1000	100	100	0.8			
1000	100	100	1.0			
1000	100	100	1.5			

TABLE B

$V_{max(1)}$	$V_{max(2)}$	F	Lowest $S_{0.5}$	v_2	$v_2/V_{max(2)}$	Sens. Index
1000	100	10	0.6			
1000	100	30	0.6			
1000	100	100	0.6			
1000	100	300	0.6			
1000	100	500	0.6			
1000	100	700	0.6			

TABLE C

$V_{max(1)}$	$V_{max(2)}$	F	Lowest $S_{0.5}$	v_2	$v_2/V_{max(2)}$	Sens. Index
500	500	10	0.6			
500	500	30	0.6			
500	500	100	0.6			
500	500	300	0.6			
500	500	500	0.6			
500	500	700	0.6			

TABLE D

$V_{max(1)}$	$V_{max(2)}$	F	Lowest $S_{0.5}$	v_2	$v_2/V_{max(2)}$	Sens. Index
500	500	100	0.1			
500	500	100	0.2			
500	500	100	0.3			
500	500	100	0.5			
500	500	100	0.6			
500	500	100	0.8			
500	500	100	1.0			
500	500	100	1.5			

Chapter 5   Experiment 5.5

dyn195E

TABLE E

$V_{max(1)}$	$V_{max(2)}$	F	Lowest $S_{0.5}$	v_2	$v_2/V_{max(2)}$	Sens. Index
1000	10	100	0.1			
1000	10	100	0.2			
1000	10	100	0.3			
1000	10	100	0.5			
1000	10	100	0.6			
1000	10	100	0.8			
1000	10	100	1.0			
1000	10	100	1.5			

dyn195F

TABLE F

$V_{max(1)}$	$V_{max(2)}$	F	Lowest $S_{0.5}$	v_2	$v_2/V_{max(2)}$	Sens. Index
1000	1	200	0.6			
1000	3	200	0.6			
1000	10	200	0.6			
1000	30	200	0.6			
1000	100	200	0.6			
1000	300	200	0.6			
1000	500	200	0.6			
1000	700	200	0.6			

Problems

The number of permutations of the four assignable parameters in the model described in Figure 5.12 is enormous. In the tables above, only one variable at a time was varied.

You can discover other effects at metabolic branch points by simultaneously changing two or more variables, either directly or reciprocally.

Conclusion

In considering the meaning of the Sensitivity Index values you obtained in this experiment, recall that variations in $S_{0.5}$ are usually caused by the binding of modifier metabolite such as the end product of the branch pathway and that the modifier usually binds cooperatively so that a small change in the concentration of the modifier may cause a relatively large change in the value of $S_{0.5}$.

Experiment 5.6*

Feedback Control of a Biosynthetic Pathway

The concentrations of intermediary metabolites and the precursors of macromolecular synthesis are stabilized — and the rates of their syntheses adjusted to meet momentary metabolic needs — by feedback control. Control systems have evolved properties that permit large changes in reaction rates and fluxes of metabolites while holding the concentrations of key intermediates within relatively narrow ranges. For example, when the rate of protein synthesis increases, feedback controls adjust the rates of amino acid synthesis so as to avoid drastic changes in amino acid pool sizes.

In this experiment you will build an electronic spreadsheet "machine" to simulate the feedback control of a metabolic pathway whose final product modifies the affinity of the regulatory enzyme at the beginning of the sequence for its substrate. The model that you will build is different from many of the others that you have experienced in this text, and its successful use may require special effort on your part. The ideas are no more difficult than those you have already successfully mastered, but the ideas (especially with respect to the model design itself) may require you to "change mental gears." Do it!

Consider a branch point off of a main pathway that begins with the metabolite B and its enzyme B-ase (Figure 5.13).

Figure 5.13

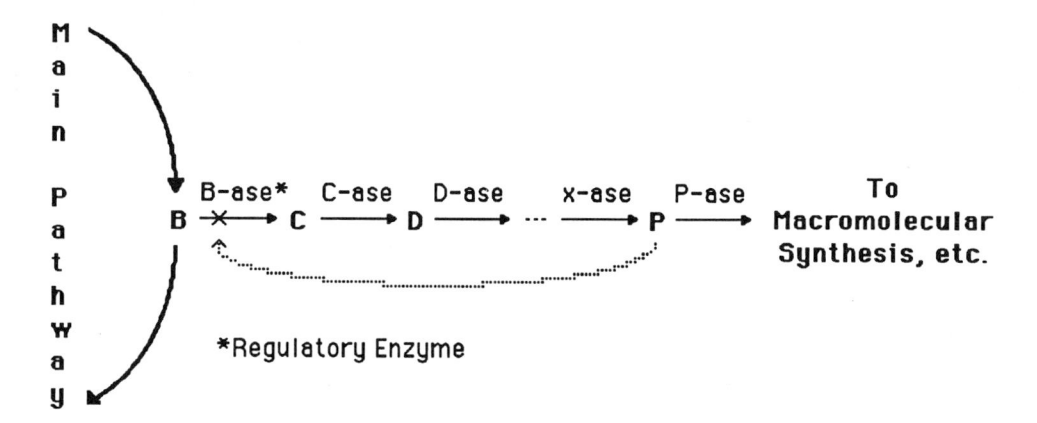

```
Main Pathway
       │
       ▼      B-ase*   C-ase    D-ase    x-ase    P-ase      To
              B ─✗─→ C ─────→ D ─────→ ··· ─────→ P ─────→ Macromolecular
                ↑                                              Synthesis, etc.
                └······················································┘

              *Regulatory Enzyme
```

Chapter 5   Experiment 5.6*

The diversion of substrate B down this side path is regulated by the concentration of end product P, many enzymic reactions away. Regulation occurs because B-ase contains modifier sites that bind end product with a resulting decrease in the affinity of the catalytic sites of the enzyme for substrate.

As the concentration of end product rises, the fraction of B-ase molecules that bind P, and thus have reduced affinity for substrate, also rises. The rate of the reaction catalyzed by B-ase therefore decreases with a consequent reduction in [P]. This negative-feedback control leads to a steady state in which flux and concentrations are constant. A change in the rate at which the end product is used can cause a quick change to a new steady state.

The mathematics that describes this feedback loop is straightforward and should already be familiar to you. Because B-ase, the regulatory enzyme at the beginning of the sequence, is rate limiting, input into the *P pool* is controlled by B-ase even if it is many reactions away. Output from the P pool is controlled by P-ase, a classical Michaelis enzyme for purposes of this model.

Consider output first. The rate of removal of P from the P pool is governed by the Michaelis equation, Equation 4.5 (Chapter 4, Introduction). Output from the P pool is summarized in Figure 5.14b.

Next consider input. To keep the model simple, assume that substrate binds to four catalytic sites on B-ase with infinite cooperativity and that end product binds to four regulatory sites on B-ase with infinite cooperativity. Two enzyme forms contribute to the rate of delivery of P into the P pool — B-ase and B-ase $\cdot P_4$. The sum of the conversions by both forms of the enzyme constitutes the total rate of input into the P pool and is summarized in Figure 5.14a. Examine the terms of this equation and confirm that they are readily derivable from the mathematics already developed in Experiment 4.7 (Equation 4.32). μ is used in the first term of this equation to express the factor by which $B_{0.5}$ is changed when the enzyme is converted from B-ase to B-ase $\cdot P_4$.

Feedback Control of a Biosynthetic Pathway

Figure 5.14

$$\text{(a)} \quad \text{B-ase} \qquad\qquad \text{(b)} \quad \text{P-ase}$$

$$\frac{[B]^4 \left(\frac{EP}{E_t}\right) V_{max}}{[\mu B_{0.5}]^4 + [B]^4} + \frac{[B]^4 \left[1 - \left(\frac{EP}{E_t}\right)\right] V_{max}}{[B]_{0.5}^4 + [B]^4} \longrightarrow P \longrightarrow \frac{V_{max}[P]}{[P] + K_m}$$

$$\text{where} \quad \frac{EP}{E_t} = \frac{[\text{B-ase} \cdot P_4]}{[\text{B-ase}]_{tot}} = \frac{[P]^4}{[P]^4 + M_{0.5}^4}$$

In the model that you will build, a feedback control "machine" has been designed around the above equations. Electronic spreadsheet cells that define the conditions of P-ase examine the current value of [P] and calculate the amount of P that leaves the P pool over a unit of time. Electronic spreadsheet cells that define the conditions of B-ase examine the same value of [P] and calculate the amount of P that enters the P pool over a unit of time. The inputs and outputs to the P pool are used to update [P] and a new value of [P] is fed into the cell that drives B-ase and P-ase. Judicious use of forward references and self-references make this possible (Experiment 2.5).

Each time the model is commanded to recalculate, it iterates one step closer to the state of the system at steady-state equilibrium. Eventually the model stabilizes around the steady-state value of the simulated feedback network. When you change the parameters that simulate a change in the demand for P (i.e., changes in the V_{max} or K_m of P-ase), further iteration causes the model to settle into a new steady state.

Remember that changes in P-ase parameters are artificial devices of your model that are used to simulate a changed demand for P. In the real world, a change in the demand for P would derive from, for example, the complex regulatory mechanisms governing protein or nucleic acid synthesis.

One final point should be mentioned. It is useful to monitor the concentration of a typical intermediary metabolite in the pathway (say, [D] in

Chapter 5   Experiment 5.6*

Figure 5.13). This can be done by computing [D] from Equation 5.31 (or Equation 5.26 in Experiment 5.4):

$$[D] = \frac{K_m v}{V_{max} - v} \quad (5.31)$$

where v is equal to the current reaction rate of B-ase.

Directions

1. Build the model described in Figure 5.15. Set the recalculation order of your electronic spreadsheet program to columnwise even if your program is able to untangle forward references.

   Cells E2, E4, and E10 contain forward references to cell E12. In some spreadsheet programs, these and other cells will show "ERROR" when the model is set up. To get the model moving replace the formula in cell E12 with a very low value (e.g., 10^{-5}). If your model is set to recalculate automatically all cells on data entry (which it should be), the model resets itself to a condition where [P] has a low value.

   Now replace the value in cell E12 with its original formula, =E12+E10. With this formula back in place, [P] is now ready to respond to changes in the actions of B-ase and P-ase.

   Use the command on your electronic spreadsheet program to repeatedly "force" recalculation until the change in [P] from one iteration to the next is less than 0.01% (cell E16).

   Except for cells H7, H8, and H11 your model should now have the same values as the model described in Figure 5.15. Cell H11 should have approximately the same value.

   Cells H7 and H8 are not tested by this verification procedure. You might want to visually inspect the contents of these cells a final time.

2. Explore the three major groups of columns: A - B (constants), D - E (engine), and G - H (memory). Remember that calculations occur in columnwise order.

Feedback Control of a Biosynthetic Pathway

3. The *constants* group of columns contains the defining parameters of the enzymes P-ase, B-ase, and D-ase. This group constitutes input to the model, and no formulas are present.

4. The *engine* group of columns contains the "machine" that drives the feedback loop.

   P(Out) (cell E2) applies the Michaelis equation to the P-ase parameters in the Constants group and the current value of [P] in cell E12. *The formula in cell E2 contains a forward reference to the value of [P] which was computed in the previous computation round.* At steady state, P(Out) is necessarily equal to the flux of metabolite through the pathway.

   P(In) (cell E7) sums up the reaction rates of the reactions catalyzed by the two different forms of B-ase (cells E5 and E6). Like P(Out), the reaction rates for B-ase are determined by using the current value of [P] in cell D12. *The formula in cell E4* calculates the fraction of B-ase that exists as B-ase • P4 and *contains a forward reference to the value of [P] which was computed in the previous computation round.* At steady state, P(In) is also a measure of the flux of metabolite through the pathway and must be equal to P(Out).

   Δ[P] (cell E9) calculates the change in [P] and P(In) – P(Out), which is about to occur.

   $[P]_{n-1}$ (cell E10) takes a snapshot of the old value of [P] (cell E12) before it is updated with ΔP (cell E9). This snapshot is used to compute the percent change in [P] (cell E16), which tells you how well your model is converging at the end of each iteration.

   $[P]_n$ (cell E12) is the value of [P] used in all calculations that went before. Now that those calculations are complete, it is time to update [P] with the new value of [P] by adding Δ[P] to it. *This cell is both the object of the forward references in cells E2 and E4 and the object of a self-reference.* Do you see how this works?

   The remainder of this column is self-explanatory.

Chapter 5   Experiment 5.6*

5. *Memory* is not an essential feature of the model, but it is a convenience that allows you to remember the values of steady-state flux and [P] under one set of conditions while iterating down to the steady state of a different set of conditions. When the toggle switch in cell H2 is set at a value of one, the cells in this column continually update. When the toggle switch in cell H2 is set at any other value (e.g., zero) the values are frozen until the toggle is reset.

Explore the formulas in these cells until you see why it works. You will receive explicit instructions on its use shortly.

6. Under some conditions (e.g., when a parameter is changed by too large a factor in one step), the model may give physically impossible answers — especially negative values of [P] that increase without limit. If your model ever journeys into this netherworld, you should reset your model according to the procedures in Directions, number 1.

7. A final note: The model implicitly assigns both volume and [B] a value of one. One consequence of this assignment is to equate amounts with concentrations, making the flow of the model easier to follow. As an exercise you may incorporate units and assignable volumes if you wish.

Feedback Control of a Biosynthetic Pathway

Figure 5.15

Formulas

	A	B	C	D	E	F	G	H
1	CONSTANTS:			ENGINE:			MEMORY:	
2	P-ase:			P.(Out) =	=(B3*(E12))/((E12)+(B4))		Update Toggle =	1
3	Vmax =	2					(1 = Yes)	
4	Km =	15		EP4/E =	=(E12*E12*E12*E12)/((E12*E12*E12*E12)+(B9*B9*B9*B9))			
5	B-ase:			v(E) =	=((1-E4)*B7)/		Prev Flux =	=IF(H2=1,H6,H5)
6				v(EP4) =	=(E4*B7)/((B10*B10*B10*B8*B8*B8)+1)		Pres Flux =	=IF(H2=1,E2,H6)
7	Vmax =	4		P. (In) =	=E5+E6		Prev P =	=IF(H2=1,H8,H7)
8	(B)0.5 =	0.5		∆[P] =	=E7-E2		Pres P =	=IF(H2=1,E12,H8)
9	(M)0.5 =	3		[P]n-1 =	=E12			
10	μ =	20					Sens Ind =	=LN(H6/H5)/LN(H8/H7)
11				[P]n =	=E12+E9			
12	D-ase:			v/Vmax(B) =	=E7/B7	<===	PRIMER	
13	Vmax =	8		[D] =	=(B14*E7)/(B13-E7)			
14	Km =	1						
15				%.Change =	=100*(E12-E10)/E10			
16								

Values

	A	B	C	D	E	F	G	H
1	CONSTANTS:			ENGINE:			MEMORY:	
2	P-ase:			P.(Out) =	0.4873		Update Toggle =	1
3	Vmax =	2					(1 = Yes)	
4	Km =	15		EP4/E =	0.8707			
5	B-ase:			v(E) =	0.4870		Prev Flux =	0.4873
6				v(EP4) =	0.0003		Pres Flux =	0.4873
7	Vmax =	4		P. (In) =	0.4873		Prev P =	4.8322
8	(B)0.5 =	0.5		∆[P] =	0.0000		Pres P =	4.8322
9	(M)0.5 =	3		[P]n-1 =	4.8322			
10	μ =	20					Sens Ind =	1.3198
11				[P]n =	4.8322	<===	PRIMER	
12	D-ase:			v/Vmax(B) =	0.1218			
13	Vmax =	8		[D] =	0.0649			
14	Km =	1						
15				%.Change =	0.0000			
16								

Recalculation Order = COLUMNWISE

Chapter 5   Experiment 5.6*

Questions

1. Use the model described in Figure 5.15 to complete the tables below. First, enter the values assigned to each of the enzyme parameters in the constants column. Second, drive the model to steady state by using the instructions given under "Directions," number 6.

   Record all values requested in the table except "sensitivity index." Sensitivity index (similar but not identical to that in Experiment 5.5) requires values obtained from two separate, closely neighboring steady states. Therefore set the update toggle in the memory column to zero (or any value other than one) and then change the value of either P-ase K_m or P-ase V_{max} by 1%. Iterate the model to a new steady state and then update memory by setting the update toggle back to one. If automatic recalculation is on (which it should be), resetting the update toggle will make available the values of flux and [P] from both the current and the previous steady states and permit the calculation of sensitivity index.

   Some portions of the table below have been completed for you so that you can verify that you are using the model correctly. The completed table is, of course, also found in the back of the book.

	V_{max}	$[S]_{0.5}$ or K_m	$[M]_{0.4}$	μ
B-ase:	4	0.5	3	20
D-ase:	8	1		
P-ase:	(variable)	6		

P-ase V_{max} =	5	4	3	2	1	0.5
B-ase v/V_{max}	0.43	__	__	__	__	__
Flux	1.71	__	__	__	__	__
[P]	3.14	__	__	__	__	__
[D]	0.27	__	__	__	__	__
Sens. Index*	-2.17	__	__	__	__	__

   *Your answers may vary slightly depending upon the increment chosen.

Feedback Control of a Biosynthetic Pathway

	V_{max}	$[S]_{0.5}$ or K_m	$[M]_{0.4}$	μ
B-ase:	4	0.5	3	20
D-ase:	8	1		
P-ase:	2	(variable)		

P-ase $K_m =$	1	2	3	5	10	15
B-ase v/V_{max}	0.38					
Flux	1.53					
[P]	3.29					
[D]	0.24					
Sens. Index	-2.36					

a. In typical biosynthetic feedback systems fluxes can vary over wide ranges whereas metabolite concentrations are held to within relatively close limits. Plot either of the above tables for both flux and [P] on the same graph and determine the range of change for both flux and [P]. Is the simulation behaving as a "proper" feedback system should?

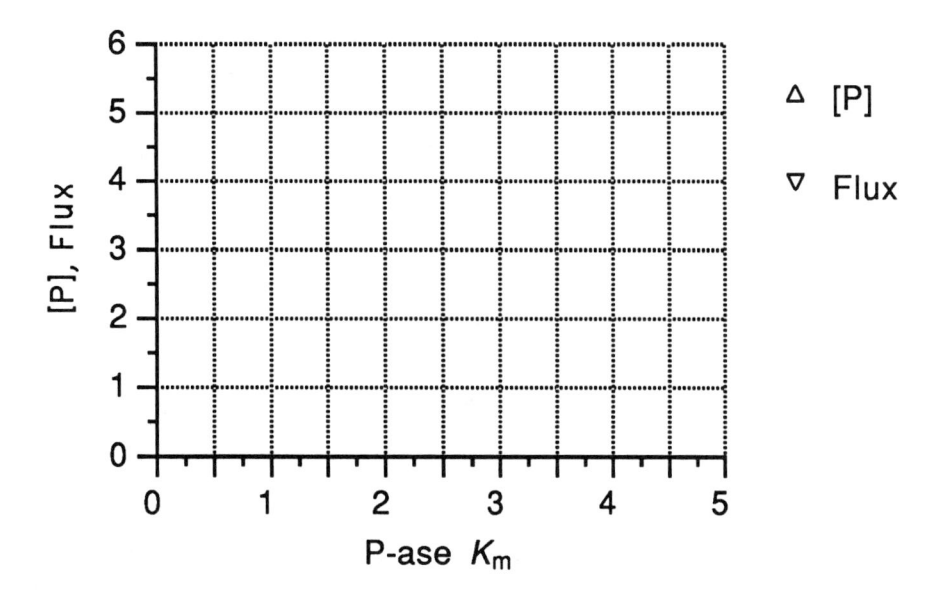

b. Now plot flux against [D]. Is the range of change of [D] significantly less than the range of change of flux?

205

Chapter 5   Experiment 5.6*

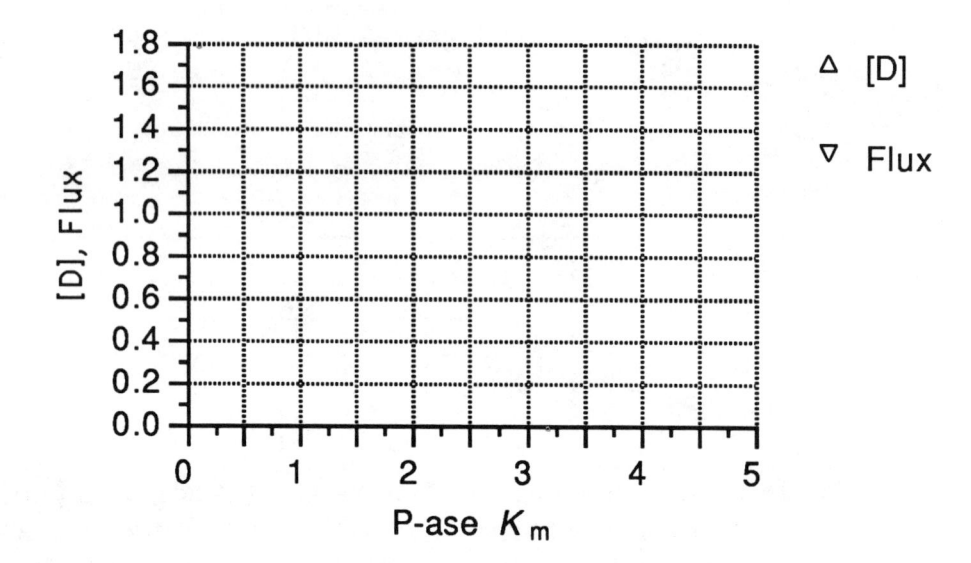

[Graph: y-axis labeled "[D], Flux" from 0.0 to 1.8; x-axis labeled "P-ase K_m" from 0 to 5. Legend: △ [D], ▽ Flux]

Feedback control of biosynthetic pathways has evolved to hold the concentrations of end products within rather narrow limits despite fluctuations in metabolic demand for them. Thus they are available at appropriate concentrations when needed by the cell. There is no need for such stabilization of concentrations of intermediates in the sequences. The interactions you explored in Experiment 5.4 determine the concentrations of such intermediates.

2. Use the data from Question 1 to compare the direction of change of [P] with flux and of [D] with flux. Can you explain the differences in behavior? (**Hint:** Recall the role that [P] plays in a feedback system and the behavior of metabolites in a pathway interior, as explored in Experiment 5.4.)

3. The sensitivity index is a convenient method for indicating how the flux through the sequence responds to changes in [P]. Thus it measures the sensitivity of the feedback-control system: sensitivity index is not calculated as in Experiment 5.5 but rather is analogous to the kinetic order of a reaction (Experiment 4.1) and is the slope of a plot of log flux as a function of log [P] (Equation 5.16). It is is obtained by dividing the difference between the logs of the fluxes at two closely spaced points by the logs of [P] at the same two points (cell H10 in Figure 5.15). (Why must the points be closely spaced?)

Feedback Control of a Biosynthetic Pathway

An index of -3, for example, would indicate that an eight-fold increase in the rate of synthesis of P would result from a decrease in [P] of only two-fold.

Examine the data in Question 1 and determine whether increases in flux lead to increases or decreases in the sensitivity of feedback control. Why must this be so?

4. Predict how changes in μ affect the sensitivity of feedback control and then design and complete a table that tests your prediction.

As μ decreases from values greater than one, does the sensitivity of control increase or decrease? What happens when $\mu = 1$?

Did you notice that convergence toward steady state became very sluggish as μ decreased? This poor convergence is itself an indication of the low sensitivity of control under these conditions.

(**Note:** In some cases, especially when μ is small, the signal parameter "% change," which is $\Delta[P]/[P]$, may initially increase with each iteration, suggesting that the model is not converging. This merely indicates that conditions are such that the fractional change of $\Delta[P]$ is less than the fractional change of $[P]$. The system is still converging as you will see as you continue your iterations.)

Problems

1. A given flux through a sequence can be obtained by many different combinations of values of V_{max} and K_m for the enzyme P-ase. Will the value of the sensitivity index be the same for all conditions that produce the same flux, or will it vary according to the properties of the enzyme P-ase?

   You can test your intuitive answer by taking an arbitrary pair of values (e.g., $V_{max} = 2$, $K_m = 15$) and determining the sensitivity index. Change the value of V_{max} and vary K_m systematically (e.g., $K_m = 10, 5, 2, 1$) until you find the value of K_m that produces the same flux as the original V_{max}, K_m pair. Are the values of the sensitivity index the same within experimental error?

Chapter 5   Experiment 5.6*

Do this for a variety of initial pairs and then provide an explanation for the generalization that your simulations have established.

2. You found in Question 3 that the sensitivity index decreases as the value of v/V_{max} for the regulatory enzyme increases. It should be obvious to you that other factors also determine the *intrinsic* control sensitivity of a sequence — number of catalytic sites, number of modifier sites, degree of cooperativity in the binding of substrate and modifier, and the value of the interaction constant μ.

You explored the effect of μ in Question 4.

Design and build models to explore the influence of other factors and influence feedback loop sensitivity.

Experiment 5.7*

Pairs of Oppositely Directed Metabolic Sequences

Metabolic conversions within living cells must be able to proceed in both directions to meet changing conditions — Glycolysis vs. gluconeogenesis, fat storage vs. fat oxidation, and amino acid synthesis vs. amino acid oxidation.

You might imagine that this requirement for bidirectionality implies a system near equilibrium in which changes in the concentrations of substrates or products are capable of changing the direction of net conversion. However,

1. Metabolism is an energy-transducing system: It makes chemical energy available for chemical conversions, movement of molecules across membranes, mechanical motion, and so forth. No energy can be obtained from a system in equilibrium.

2. Metabolism is adaptable and responsive to changing conditions. Conversion rates and directions respond sensitively to the momentary needs of the cell or organism for each conversion. Such sensitivity is impossible in a system near equilibrium because systems near equilibrium are likely to be kinetically sluggish and the direction of conversion depends on only thermodynamic considerations (the direction toward equilibrium).

It is impossible to design a responsive and adaptive metabolic system in which, for example, the conversions of C to D and D to C are carried out by the same chemical reaction in opposite directions. The only way to ensure that both directions of conversion are always available to the cell on demand is to provide each conversion direction with its own set of enzymes and its own set of regulatory signals. In addition, the overall equilibrium constants for the two directions must be different — a requirement that demands that the overall stoichiometries of the two directions must be different.

In metabolism, oppositely directed metabolic conversions are usually made possible by differences in the numbers of ATP molecules that are hydrolyzed or regenerated in the two pathways.

Chapter 5   Experiment 5.7*

Consider the simplest case of two such reactions (Figure 5.16). These reactions closely resemble the pair of reactions catalyzed by phosphofructokinase and fructose-1,6-bisphosphate phosphatase, where P_i is phosphate ion (Figure 5.17).

In the case of the phosphatase reaction (reaction II, Figure 5.17), the equilibrium ratio of fructose-6-phosphate to fructose-1,6-bisphosphate under physiologic conditions is about 10,000.

Figure 5.16

(I)   C + ATP ⇌ D + ADP

(II)   D ⇌ C

Figure 5.17

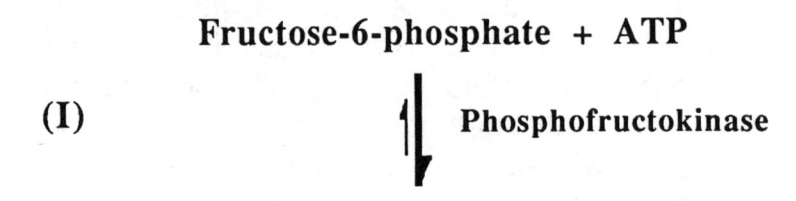

(I)

Fructose-6-phosphate + ATP
⇅ Phosphofructokinase
Fructose-1,6-bisphosphate + ADP

(II)

Fructose-1,6-bisphosphate
⇅ Phosphatase
Fructose-6-phosphate + P_i

Figure 5.18

		1st Order	2nd Order
(I)	C ⇌ D (ATP → ADP)	$\dfrac{V_{max}[C]}{K_c + [C]}$	$\dfrac{V_{max}[C]^2}{K_c^2 + [C]^2}$
(II)	D ⇌ C	$\dfrac{V_{max}[D]}{K_d + [D]}$	$\dfrac{V_{max}[D]^2}{K_d^2 + [D]^2}$

Where K_c and K_d are the K_m's of their respective reactions

In the case of the phosphofructokinase reaction (reaction I, Figure 5.17), the equilibrium ratio of fructose-1,6-bisphosphate to fructose-6-phosphate under physiologic conditions is also about 10,000 because the reaction is coupled to the conversion of ATP to ADP.

You will consider the implications of the preceding discussion with two separate models. In one case, first-order Michaelis-type enzymes are considered. In the other case, second-order Michaelis-type enzymes are considered. You are already familiar with the mathematics that support these models, which are summarized in Figure 5.18. In each case, V_{max} of the reverse reaction is approximated by use of the expression:

$$K_{eq} = \frac{V_{max(fwd)} K_{prod}}{V_{max(rev)} K_{sub}} \qquad (5.32)$$

where $V_{max(fwd)}$ = V_{max}, forward reaction
$V_{max(rev)}$ = V_{max}, reverse reaction
K_{prod} = product dissociation constant
K_{sub} = substrate dissociation constant

Directions

1. Build and verify the models described in Figures 5.19 and 5.20.

2. In the case of both models, [C], [D], K_{eq}, V_{max} (forward reaction), and the dissociation constants (from enzyme) of both C and D are inputs into the model.

3. V_{max} (reverse reaction) is computed from Equation 5.32.

4. The net conversions C → D and D → C, as well as their difference, are computed from the equations in Figure 5.18.

5. "% waste" calculates the rate of reaction in the direction opposite to the net conversion divided by the net flux. This value measures the percent of the net conversion that is wasted by back-conversion, and its significance will become clear as the experiment proceeds.

Pairs of Oppositely Directed Metabolic Sequences

Figure 5.19

Formulas

	A	B	C	D	E
1	1st Order Reactions:				
2	[C] =	0.1		v(I,fwd) =	=(B6*B2)/(B7+B2)
3	[D] =	0.2		v(I,rev) =	=(B15*B3)/(B7+B3)
4	RXN I: C + ATP --> D + ADP			v(II,fwd) =	=(B11*B3)/(B3+B12)
5	Keq =	10000		v(II,rev) =	=(B16*B2)/(B13+B2)
6	Vmax =	10			
7	Kc =	0.1		v(C-->D) =	=E2+E5
8	Kd =	1		v(D-->C) =	=E4+E3
9	RXN II: D --> C				
10	Keq =	20000		v(net) =	=E7-E8
11	Vmax =	20			
12	Kd =	0.2		% "waste" =	=100*(MIN(E7..E8)/ABS(E10))
13	Kc =	2			
14					
15	RXN I: Vm(rev) =	=(B6*B8)/(B5*B7)			
16	RXN II: Vm(rev) =	=(B11*B13)/(B12*B10)			

Values

	A	B	C	D	E
1	1st Order Reactions:				
2	[C] =	0.1		v(I,fwd) =	5.0000
3	[D] =	0.2		v(I,rev) =	0.0067
4	RXN I: C + ATP --> D + ADP			v(II,fwd) =	10.0000
5	Keq =	10000		v(II,rev) =	0.0005
6	Vmax =	10			
7	Kc =	0.1		v(C-->D) =	5.0005
8	Kd =	1		v(D-->C) =	10.0067
9	RXN II: D --> C				
10	Keq =	20000		v(net) =	-5.0062
11	Vmax =	20			
12	Kd =	0.2		% "waste" =	99.8859
13	Kc =	2			
14					
15	RXN I: Vm(rev) =	0.01			
16	RXN II: Vm(rev) =	0.01			

Chapter 5   Experiment 5.7*

Figure 5.20

Formulas

	A	B	C	D	E
1	2nd Order Reactions:				
2	[C] =	0.3		v(a,fwd) =	=(B6*B2*B2)/((B7*B7)+(B2*B2))
3	[D] =	0.3		v(a,rev) =	=(B15*B3*B3)/((B8*B8)+(B3*B3))
4	RXN I: C + ATP --> D + ADP			v(b,fwd) =	=(B11*B3*B3)/((B3*B3)+(B12*B12))
5	Keq =	10000		v(b,rev) =	=(B16*B2*B2)/((B13*B13)+(B2*B2))
6	Vmax =	10			
7	Kc =	0.3		v(C-->D) =	=E2+E5
8	Kd =	1		v(D-->C) =	=E4+E3
9	RXN II: D --> C				
10	Keq =	10000		v(net) =	=E7-E8
11	Vmax =	10			
12	Kd =	3		% "waste" =	=100*(MIN(E7..E8)/ABS(E10))
13	Kc =	1			
14					
15	RXN I: Vm(rev) =	=(B6*B8*B8)/(B5*B7*B7)			
16	RXN II: Vm(rev) =	=(B11*B13*B13)/(B12*B12*B10)			

Values

	A	B	C	D	E
1	2nd Order Reactions:				
2	[C] =	0.3		v(a,fwd) =	5.0000
3	[D] =	0.3		v(a,rev) =	0.0009
4	RXN I: C + ATP --> D + ADP			v(b,fwd) =	0.0990
5	Keq =	10000		v(b,rev) =	0.0000
6	Vmax =	10			
7	Kc =	0.3		v(C-->D) =	5.0000
8	Kd =	1		v(D-->C) =	0.0999
9	RXN II: D --> C				
10	Keq =	10000		v(net) =	4.9001
11	Vmax =	10			
12	Kd =	3		% "waste" =	2.0393
13	Kc =	1			
14					
15	RXN I: Vm(rev) =	0.0111			
16	RXN II: Vm(rev) =	0.0001			

Questions

1. In most paired conversions, the kinetic controls on the two oppositely directed sequences act in a push-pull manner; that is, the affinity for substrate of the first enzyme of one sequence decreases as the affinity for substrate of the first enzyme of the oppositely directed sequence increases. This feature is imposed on the table

Pairs of Oppositely Directed Metabolic Sequences

that follows. Complete the table for first-order enzymes with K_{eq} values of 10,000 and V_{max} values of 10 (Figure 5.19). This situation mimics closely the phosphofructokinase–phosphatase sequence pair just discussed.

(**Note:** Because all enzyme parameters except K_c (reaction I) and K_d (reaction II) are identical, your results will be symmetrical around $K_c = K_d = 1$. This is a convenient setting for exercising the model but is not a necessary feature of oppositely directed sequences.)

[C]	[D]	K_c (RXN I)	K_d (RXN II)	$v(C \rightarrow D)$	$v(D \rightarrow C)$	v_{net}	% Waste
1	1	10	0.1	___	___	___	___
1	1	5	0.2	___	___	___	___
1	1	2.5	0.4	___	___	___	___
1	1	1	1	___	___	___	___
1	1	0.4	2.5	___	___	___	___
1	1	0.2	5	___	___	___	___
1	1	0.1	10	___	___	___	___
0.3	0.3	1	1	___	___	___	___
0.3	0.3	0.4	2.5	___	___	___	___
0.3	0.3	0.2	5	___	___	___	___
0.3	0.3	0.1	10	___	___	___	___
0.1	0.1	1	1	___	___	___	___
0.1	0.1	0.4	2.5	___	___	___	___
0.1	0.1	0.2	5	___	___	___	___
0.1	0.1	0.1	10	___	___	___	___

$K_d(\text{RXN I}) = K_c(\text{RXN II}) = 1$

a. The net conversion is, of course, zero, and the % waste is infinitely high when the rates of the two reactions are equal. In this table, because we have assigned equal values of K_{eq} and V_{max} for the two enzymes and equal concentrations for C and D, the rates are equal when K_c (RXN I) equals K_d (RXN II).

This condition, if it existed in nature, would lead only to wasteful hydrolysis of ATP with no metabolic consequences. The term *futile-cycling* is sometimes used to describe this hypothetical situation.

b. The net conversion increases and the % waste decreases as the kinetic properties of the enzymes become widely different: for

Chapter 5   Experiment 5.7*

example, when K_c (RXN I) is very different in value from K_d (RXN II). Explain why this is so.

c. What happens to % waste as the concentrations of substrates are reduced? Would the consequences be the same if [C] and [D] remained constant and the values of K_c and K_d had increased?

It is important to remember that the crucial parameter driving reaction rates is $[S]/K_s$ (Equation 4.6).

2. Complete the following table for a second-order enzyme (Figure 5.20). Conditions are identical to those in Question 1.

[C]	[D]	K_c (RXN I)	K_d (RXN II)	$v(C \to D)$	$v(D \to C)$	v_{net}	% Waste
1	1	10	0.1				
1	1	5	0.2				
1	1	2.5	0.4				
1	1	1	1				
1	1	0.4	2.5				
1	1	0.2	5				
1	1	0.1	10				
0.3	0.3	1	1				
0.3	0.3	0.4	2.5				
0.3	0.3	0.2	5				
0.3	0.3	0.1	10				
0.1	0.1	1	1				
0.1	0.1	0.4	2.5				
0.1	0.1	0.2	5				
0.1	0.1	0.1	10				

K_d(RXN I) = K_c (RXN II) = 1

a. Examine the table (plot portions of it, if you wish) and verify that the pattern of response is similar to that in Question 1 but that the introduction of cooperative kinetics has sharpened the regulatory properties of the system. For any pair of concentrations, the net flux is somewhat larger and the % waste is much lower for second-order reactions than for first-order reactions.

Pairs of Oppositely Directed Metabolic Sequences

b. Set the K_{eq} for both reactions to one and repeat the calculations for portions of any of the preceding tables. What do your results tell you?

You saw that when there is no difference between the enzymes in kinetic properties, no control and no net flux (and therefore no metabolism) are possible. What similar statement can you make with regard to the thermodynamic differences between the sequences?

Do you see why ATP is metabolically important?

3. Complete the following sentence:

When the reactions catalyzed by the regulatory enzymes are second-order, a change of 100-fold in the value of $S_{0.5}$ or apparent K_m of each enzyme (corresponding to a change of 2.8 kcal/mol in the binding energy for substrate) causes the system to change from a net flux of _____% of V_{max} in one direction with a waste of _____% to a net flux of _____% of V_{max} in the other direction with a waste of _____%.

This statement is worth thinking about. It illustrates the features that underlie all metabolic stabilization, correlation, and adaptation.*

---

* This type of organization, with each conversion being associated with a pair of oppositely directed sequences of reactions that differ in stoichiometry, Gibbs free energy, and kinetic control, is found throughout metabolism and is essential for metabolic function and correlation in all of its aspects. Unfortunately, it is not covered explicitly in any of the standard texts. A general discussion may be found in *Cellular Energy Metabolism and its Regulation*, D. Atkinson, Academic Press, 1977.

Chapter 5   Experiment 5.7*

Problems

You might enjoy cultivating a serious interest in metabolism by modifying the models in this experiment or building similar ones. Such models can illustrate a wide range of metabolic interrelationships.

Among the many kinds of changes to the model that might be explored are:

a. Change either or both enzymes in the two sequences from two to four interacting catalytic sites.

b. Introduce simulation of the effects of modifiers (notably the adenylate energy charge or the ATP/ADP ratio) to change the affinities of the enzymes, rather than merely assigning them manually in the models described in Figures 5.19 and 5.20.

c. Introduce formulas to simulate changes in concentrations of [C] and [D] caused by changes in K_c and K_d. This should be done with care, bearing in mind that these metabolites are also substrates for other enzymes and so their concentrations depend on complex interactions between the properties of several enzymes and the fluxes through several sequences.

d. Simulate a real pair of sequences by using as closely as possible the actual properties of their initial enzymes and the concentration ranges of the key substrates. Remember, however, that although such explicit simulation can be interesting and informative, there is no instance in which all information needed to build quantitatively valid models is known. However, such models can be useful in elucidating types of interactions and ranges of consequences.

Experiment 5.8*

Redox Reactions in Metabolism

Most of the metabolic energy used by aerobic heterotrophic organisms, including you, is derived from the oxidation of substrates. Electrons are transferred from a substrate to a primary oxidizing agent, usually NAD⁺ or a flavoprotein. ATP is then regenerated from ADP as the electrons move along the electron transfer chain from the primary oxidizing agent to O_2.

The energy potentially available from this process is proportional to the number of electrons transferred and the voltage difference across which they flow (Equation 5.33).

$$\Delta G = -n\mathcal{F} \Delta \mathcal{E} \tag{5.33}$$

where ΔG = Gibbs free energy (kcal/mol)
n = number of transferred electrons
$\mathcal{F}$ = Faraday (23 kcal/volt mol)
$\Delta \mathcal{E}$ = potential difference (volts)

The minus sign on the right hand side of the equation is a consequence of arbitrary conventions with respect to Gibbs free energy and the sign of the charge of an electron.

Thus when $n = 2$, as for most metabolic primary oxidations, the change in Gibbs free energy in kcal/mol is equal to the potential difference in volts times -46.

Under standard conditions, Equations 5.1 and 5.33 can be used to obtain Equation 5.34 and, by rearrangement (and at 25°C), Equation 5.35.

$$\Delta G° = -n\mathcal{F}\Delta \mathcal{E}° = -2.3\, RT \log(K_{eq}) \tag{5.34}$$

$$K_{eq} = 10^{\left(\frac{n\mathcal{F}\Delta \mathcal{E}°}{2.3RT}\right)} = 10^{16.6n\Delta \mathcal{E}°} \tag{5.35}$$

These relationships permit the calculation of equilibrium constants for specific reactions from tables of electrochemical potentials (Table 5.1). In

Chapter 5   Experiment 5.8*

Table 5.1

Reaction	Electrode Potential
3-P-glyceraldehyde + H_2O --> 3-P-glycerate$^-$ + $3H^+$ + $2e^-$	-0.55
pyruvate$^-$ + H_2O --> acetate$^-$ + CO_2 + $2H^+$ + $2e^-$	-0.60
isocitrate^{--} + H_2O --> α-ketoglutarate^{--} + CO_2 + H^+ + $2e^-$	-0.38
α-ketoglutarate^{--} + H_2O --> succinate^{--} + CO_2 + $2H^+$ + $2e^-$	-0.67
succinate^{--} --> fumarate^{--} + $2H^+$ + $2e^-$	0.03
malate^{--} --> oxaloacetate^{--} + $2H^+$ + $2e^-$	-0.17
lactate$^-$ --> pyruvate$^-$ + $2H^+$ + $2e^-$	-0.19
pyruvate$^-$ + HSCoA --> AcSCoA + CO_2 + H^+ + $2e^-$	-0.51
ethanol --> acetaldehyde + $2H^+$ + $2e^-$	-0.20
acetaldehyde + HSCoA --> AcSCoA + $2H^+$ + $2e^-$	-0.41
NADH --> NAD^+ + H^+ + $2e^-$	-0.34

this experiment you will use potentials that are calculated for pH 7 according to the convention that an algebraically more positive potential corresponds to a stronger oxidizing agent.

Directions

1. Build and verify the model described in Figure 5.21.

2. The model assumes that $\mathfrak{F}$ = 23 kcal/volt mole and is therefore designed to accept input of $\mathcal{E}°$ in volts and to calculate $\Delta G°$ in kcal/mol. If you prefer kjoules/mol, the factor of 23 in the cells of column F should be 96.5.

Figure 5.21

Formulas

	A	B	C	D	E	F
1				Oxid. Agent	Keq	ΔGo (kcal/mol)
2	Eo(Sub) =	-0.17		NAD	=10^(16.6*B8*(B4-B2))	=23*B8*(B2-B4)
3				FP	=10^(16.6*B8*(B5-B2))	=23*B8*(B2-B5)
4	Eo(NAD) =	-0.34		O2	=10^(16.6*B8*(B6-B2))	=23*B8*(B2-B6)
5	Eo(FP) =	0				
6	Eo(O2) =	0.82				
7						
8	n =	2				

Values

	A	B	C	D	E	F
1				Oxid. Agent	Keq	ΔGo (kcal/mol)
2	Eo(Sub) =	-0.17		NAD	2.27E-6	7.82
3				FP	4.41E5	-7.82
4	Eo(NAD) =	-0.34		O2	7.38E32	-45.54
5	Eo(FP) =	0				
6	Eo(O2) =	0.82				
7						
8	n =	2				

Questions

1. For primary oxidation reactions in metabolism the number of transferred electrons is typically two, and the change in Gibbs free energy is proportional to the difference between the electrode potential of the substrate and the electrode potential of an electron acceptor such as NAD+ or a flavoprotein. Complete and graph the following table to confirm this simple relationship.

Chapter 5   Experiment 5.8*

Substrate	NAD$^+$		FP		O$_2$	
$E°$	K_{eq}	$\Delta G°$	K_{eq}	$\Delta G°$	K_{eq}	$\Delta G°$
-0.6	___	___	___	___	___	___
-0.4	___	___	___	___	___	___
-0.2	___	___	___	___	___	___
0	___	___	___	___	___	___
0.2	___	___	___	___	___	___
0.4	___	___	___	___	___	___
0.6	___	___	___	___	___	___

2. The table below rephrases Question 1 in terms of several important metabolic reactions. Complete the table and answer the questions that follow. The relevant substrate electrode potentials may be found in Table 5.1.

	Substrate	NAD+ K_{eq}	NAD+ $\Delta G°$	FP K_{eq}	FP $\Delta G°$	O_2 K_{eq}	O_2 ΔG
1.	3-P-Glyceraldehyde						
2.	Pyruvate						
3.	Isocitrate						
4.	α-Ketoglutarate						
5.	Succinate						
6.	Malate						
7.	Lactate						
8.	Pyruvate (to AcSCoA)						
9.	Ethanol						
10.	Acetaldehyde						

a. Which three conversions in the table have values of K_{eq} of 10^7 or greater when NAD+ is the oxidizing agent?

Each of these three conversions leads to the production of an activated metabolite: (1) The conversion of 3-P-glyceraldehyde to 3-P-glycerate is mediated in two steps by 3-P-glyceraldehyde dehydrogenase and 3-P-glycerate kinase and generates a molecule of ATP; (4) the conversion of α-ketoglutarate to succinate (via α-ketoglutarate dehydrogenase and succinate thiokinase) yields a molecule of either GTP or ATP; (2) the metabolic oxidation of pyruvate actually leads to the activated intermediate acetyl-CoA (reaction 8), rather than to acetate.

It is the large values of K_{eq} (or large drops in free energy) in these conversions that makes energy coupling possible at these points.

Conversion (1) appears to be the only ATP-linked conversion that is important in both directions in vivo (toward 3-P-glycerate in glycolysis and toward 3-P-glyceraldehyde in gluconeogenesis). Examine the relative ΔG values for the three conversions (1), (2), and (4). Why do you suppose reversible ATP regeneration was more likely to evolve for conversion (1) than for either of the others?

Chapter 5   Experiment 5.8*

b. Conversion (5) is another special case. The conversion of succinate to fumarate is the only oxidative step in carbohydrate metabolism for which NAD+ is not the primary oxidizing agent. Examine your values of K_{eq} and ΔG for conversion (5) when NAD+ is the primary oxidizing agent and when flavoprotein is the primary oxidizing agent. Can you suggest why metabolism has evolved so as to use flavoprotein in this instance?

c. The value of K_{eq} for the conversion of malate to oxaloacetate (conversion [6]) with NAD+ is very small and is near the limit of thermodynamic unfavorability to be found among metabolic reactions. What factors might determine whether an equilibrium constant of this magnitude is acceptable?

In this case the K_m of citrate synthetase for oxaloacetate is very low, and the K_{eq} for the citrate synthetase reaction is large. Thus the low equilibrium ratio of oxaloacetate to malate does not seriously impede flux through the citrate cycle.

d. Both directions of the interconversion of lactate and pyruvate (conversion [7]) are important in metabolism.

Lactate is an excellent substrate for most aerobic microorganisms, but lactate is made nearly quantitatively from carbohydrate by some anaerobic bacteria. The muscles of mammals produce lactate under conditions of energy and oxygen stress, and the muscles and liver convert lactate back to carbohydrate after the stress ends. Because the conversion of pyruvate to lactate is highly favorable, why is lactate a good energy substrate? (**Hint:** Note the values of K_{eq} and ΔG for the conversion of pyruvate to acetyl-CoA in the table. Also, the ratio of [NAD+]/[NADH] is kept well above one in vivo, and the highly activated acetyl-CoA can be metabolically effective at low concentrations.)

Redox Reactions in Metabolism

Problems

1. Build and verify the model at the end of this question. The model is actually two models — both using the same input values of electrode potentials and number of electrons transferred.

   Consider the generalized oxidation-reduction reaction:

   $$A_{ox} + B_{red} \rightleftharpoons A_{red} + B_{ox}$$

   In the living cell the ratio of the oxidized to the reduced form of each electron carrier is usually held at a relatively constant ratio. Use the model region labeled "Model I" to explore how the ratio of A_{ox}/A_{red} influences the ratio of B_{ox}/B_{red} at equilibrium.

   The model region labeled "Model II" is more general and calculates the equilibrium concentrations of reactants and products from the initial concentrations of reactants and products. You can find an example of the method used to develop the model by examining Experiment 3.1.

   Formulas

	A	B	C	D	E
1	Eo(A) =		-0.34		
2	Eo(B) =		-0.39	a =	=B5-1
3	n =		2	b =	=-(B5*(B12+B13)+B14+B15)
4				c =	=B5*B12*B13-(B14*B15)
5	Keq =	=10^(16.6*B3*(B1-B2))		x =	=(-E3-SQRT(E3*E3-(4*E2*E4)))/2*E2)
6	ΔGo =	=23*B3*(B2-B1)			
7					
8	Model I:				
9	Aox/Ared =		5	Box/Bred =	=B5*B9
10					
11	Model II:				
12	Aox(init) =		5	Aox(eq) =	=B12-E5
13	Bred(init) =		2	Bred(eq) =	=B13-E5
14	Ared(init) =		1	Ared(eq) =	=B14+E5
15	Box(init) =		3	Box(eq) =	=B15+E5

Chapter 5   Experiment 5.8*

Values

	A	B	C	D	E
1	Eo(A) =		-0.34		
2	Eo(B) =		-0.39	a =	44.71
3	n =		2	b =	-323.96
4				c =	454.09
5	Keq =		45.71	x =	1.90
6	ΔGo =		-2.30		
7					
8	Model I:				
9	Aox/Ared =		5	Box/Bred =	228.54
10					
11	Model II:				
12	Aox(init) =		5	Aox(eq) =	3.10
13	Bred(init) =		2	Bred(eq) =	0.10
14	Ared(init) =		1	Ared(eq) =	2.90
15	Box(init) =		3	Box(eq) =	4.90

Conclusion

The $E°$ values for other compounds that are oxidized or reduced in metabolism can be found in tables in reviews, textbooks, and monographs. However, if you are not familiar with the muddled world of electrochemical conventions, you must be on your guard when using these tables.

Formerly there were two sign conventions, and some older tables have all signs reversed from those noted in Table 5.1. You can check which convention is being used by noting the signs of strongly reducing couples such as Li/Li+ or Na/Na+. These potentials would be negative in the convention chosen for Table 5.1. Using this currently universal convention, a strongly oxidizing couple such as H_2O/O_2 has a positive electrode potential.

A more common source of possible error derives from the fact that the tables in physical chemistry handbooks and texts are based on a standard state of pH = 0, and the tables in most biochemistry handbooks and texts are based on a standard state of pH = 7. Conversions that do not liberate or consume protons (e.g., Fe^{++}/Fe^{+++}), have the same potential values in tables of both types. However, potentials for most systems of biologic interest will have different values in the two types of table.

Chapter 6

Membrane Transport

Living cells are bounded by membranes that restrict the flow of molecules into and out of the interior. In procaryotic cells these membrane structures are generally based on phospholipid bilayers; in eucaryotic cells, plasma membranes are generally based on phospholipid-cholesterol bilayers.

Both types of lipid structures have limited permeability to low molecular weight, neutral molecules (e.g., H_2O, O_2) and to hydrophobic molecules that are soluble in the membrane interior (Figure 6.1). Both types of lipid structures are essentially impermeable to hydrophilic molecules, especially polar or charged metabolites and macromolecules. Hydrophilic molecules must be transported across the membrane by membrane proteins that intercalate the lipid bilayer. These proteins usually show a high degree of specificity with respect to the molecules they are able to transport (Figure 6.1). The specificity of transport of many nutrients and waste products is therefore largely a function of the protein composition of the cell membrane.

Figure 6.1

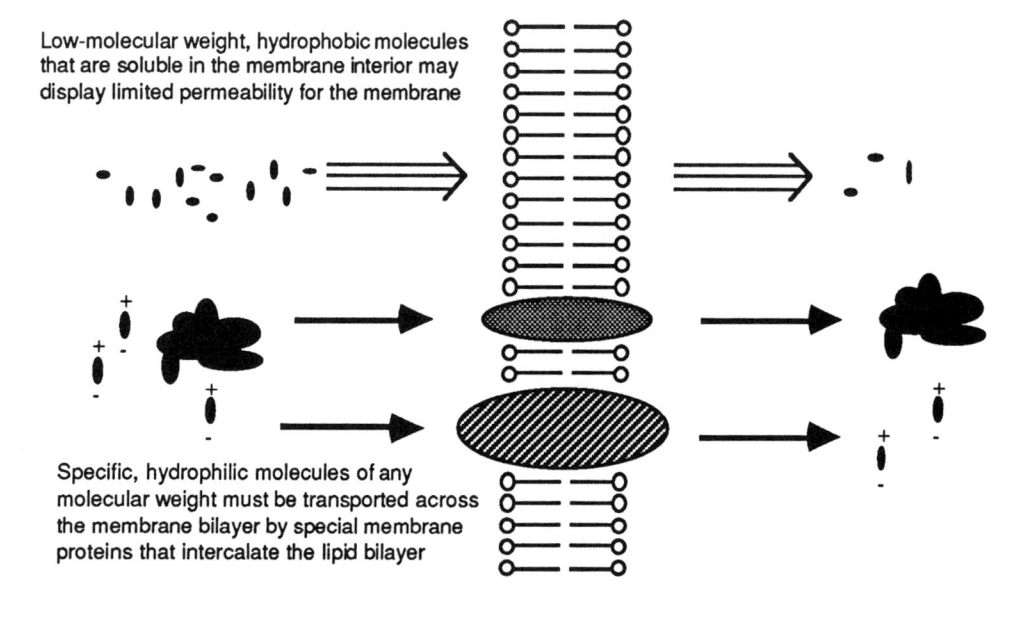

Low-molecular weight, hydrophobic molecules that are soluble in the membrane interior may display limited permeability for the membrane

Specific, hydrophilic molecules of any molecular weight must be transported across the membrane bilayer by special membrane proteins that intercalate the lipid bilayer

Chapter 6

Membranes with similar structures and transport functions are also found in the interior of eucaryotic cells where they define structures such as the endoplasmic reticulum and act as boundaries around such subcellular structures as the nucleus, lysosome, and peroxisome. Some subcellular organelles, such as mitochondria or chloroplasts, consist of multiple, concentric membrane systems.

In this chapter you will explore the quantitative aspects of several different forms of membrane transport:

1. Passive diffusion
2. Facilitated diffusion
3. Active transport

Transport routes that use membrane rearrangements (e.g., the fusion of membrane vesicles involved in neurotransmitter release) are not considered.

Although calculus is used to derive the necessary flux equations, students not familiar with this mathematics can proceed directly to the final mathematical models without difficulty.

Experiment 6.1

Passive Transport through Membranes

Consider a membrane bilayer separating two aqueous phases. The transport rate of a solute through the membrane is a function of the physical and chemical properties of the membrane itself (characterized by a permeability coefficient) and the concentration of the solute to be transported.

If: J = rate of transport across the membrane
(mmols/s/cm²)

P_s = permeability coefficient
(cm/s)

[S] = concentration of solute to be transported
(mols/L = mmols/cm³)

Then

$$J = P_s[S] \tag{6.1}$$

Consider a cell of volume V_{cell} (cm³) surrounded by a medium of effectively infinite volume (Figure 6.2).

Figure 6.2

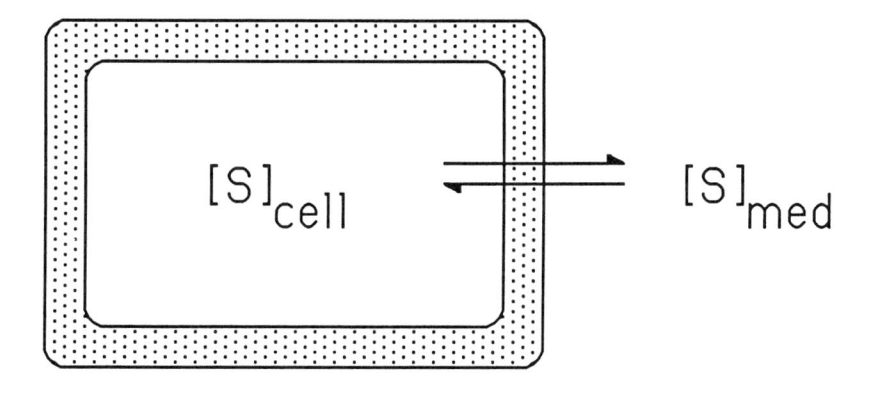

Chapter 6   Experiment 6.1

The intracellular concentration of S, $[S]_{cell}$, can be determined as follows. The rate of transport of the solute S from the medium to the cell $J_{med \to cell}$ is (from Equation 6.1):

$$J_{med \to cell} = P_s [S]_{med} \tag{6.2}$$

The rate of transport of the solute S from the cell to the medium $J_{cell \to med}$ is (from Equation 6.1):

$$J_{cell \to med} = P_s [S]_{cell} \tag{6.3}$$

The net flux of S from the medium to the cell J_{net} is then:

$$J_{net} = J_{med \to cell} - J_{cell \to med} \tag{6.4}$$

Substituting Equations 6.2 and 6.3 into Equation 6.4 yields:

$$J_{net} = P_s ([S]_{med} - [S]_{cell}) \tag{6.5}$$

The rate of change of $[S]_{cell}$, $d[S]_{cell}/dt$, can be determined by multiplying the net flux of S by the membrane surface area divided by the volume of the cell:

$$\frac{d[S]_{cell}}{dt} = \frac{(J_{net})(A)}{V_{cell}} \tag{6.6}$$

where A = membrane surface area (cm^2)
V_{cell} = cell volume (cm^3)

Substituting Equation 6.5 into 6.6 yields:

$$\frac{d[S]_{cell}}{dt} = \frac{P_s A}{V_{cell}} ([S]_{med} - [S]_{cell}) \tag{6.7}$$

or

$$\frac{d[S]_{cell}}{[S]_{med}-[S]_{cell}} = \frac{P_s A}{V_{cell}} dt \quad (6.8)$$

Because the volume of the medium is assumed to be infinite, $[S]_{med}$ is a constant and Equation 6.7 can be integrated to yield Equation 6.9 ($[S]_{med}$ is also a constant under conditions where physiologic homeostasis maintains $[S]$ at a constant level):

$$-\ln([S]_{med}-[S]_{cell}) = \frac{P_s At}{V_{cell}} + C \quad (6.9)$$

Where C is the constant of integration.

When $t = 0$, $[S]_{cell} = 0$. Therefore $C = -\ln[S]_{med}$.

Thus

$$\ln\frac{[S]_{med}}{[S]_{med}-[S]_{cell}} = \frac{P_s A}{V_{cell}} t \quad (6.10)$$

Solving for $[S]_{cell}$ yields

$$[S]_{cell} = \frac{[S]_{med} e^{\left(\frac{P_s At}{V_{cell}}\right)} - [S]_{med}}{e^{\left(\frac{P_s At}{V_{cell}}\right)}} \quad (6.11)$$

Equation 6.11 describes the transport of many small nonpolar molecules, including water and urea, across cell membranes.

Chapter 6   Experiment 6.1

Directions

1. Build and verify the model described in Figure 6.3. The values of V_{cell} and area, which initialize the model, are those of human erythrocytes.

2. The model calculates both $[S]_{cell}$ and the minimum area of membrane possible, area(min), given V_{cell} (the area surrounding a sphere of volume V_{cell}). The actual area used to calculate $[S]_{cell}$, however, is taken from cell B3.

2. Many electronic spreadsheet programs have a built-in function to accurately represent the value of π. Cells B7 and B8 of the model make use of such a built-in function.

Figure 6.3

Formulas

	A	B
1	Ps =	1.00E-7
2	Vcell =	9.70E-11
3	Area =	1.42E-6
4	[S]med =	1
5	t =	200
6		
7	Area (min) =	=(4*PI)*(((3*B2)/(4*PI))^(2/3))
8	[S]cell =	=(B4*EXP(B1*B3*B5/B2)-B4)/EXP(B1*B3*B5/B2)

Values

	A	B
1	Ps =	1.00E-7
2	Vcell =	9.70E-11
3	Area =	1.42E-6
4	[S]med =	1
5	t =	200
6		
7	Area (min) =	1.02E-6
8	[S]cell =	0.2538

Passive Transport through Membranes

Questions

1. Assume that in three separate experiments, human red blood cells (RBCs) are incubated in $1M$ solutions of either acetamide ($P_s = 4.6 \times 10^{-5}$ cm/s), butyramide ($P_s = 1.1 \times 10^{-5}$ cm/s), or glycerol ($P_s = 2.1 \times 10^{-8}$ cm/s). The volume of each RBC is 9.7×10^{-11} cm^3, and the surface area of each red cell is 1.42×10^{-6} cm^2.

   Use the model described in Figure 6.3 to follow the changes with time in the intracellular concentration of each of the substrates.

Time (s)	Acetamide [S]$_{cell}$	Butyramide [S]$_{cell}$	Glycerol [S]$_{cell}$
0	___	___	___
5	___	___	___
10	___	___	___
15	___	___	___
20	___	___	___
25	___	___	___
100	___	___	___
300	___	___	___
500	___	___	___
700	___	___	___
900	___	___	___
1000	___	___	___
10000	___	___	___

△ Acetamide   ▽ Butyramide   □ Glycerol

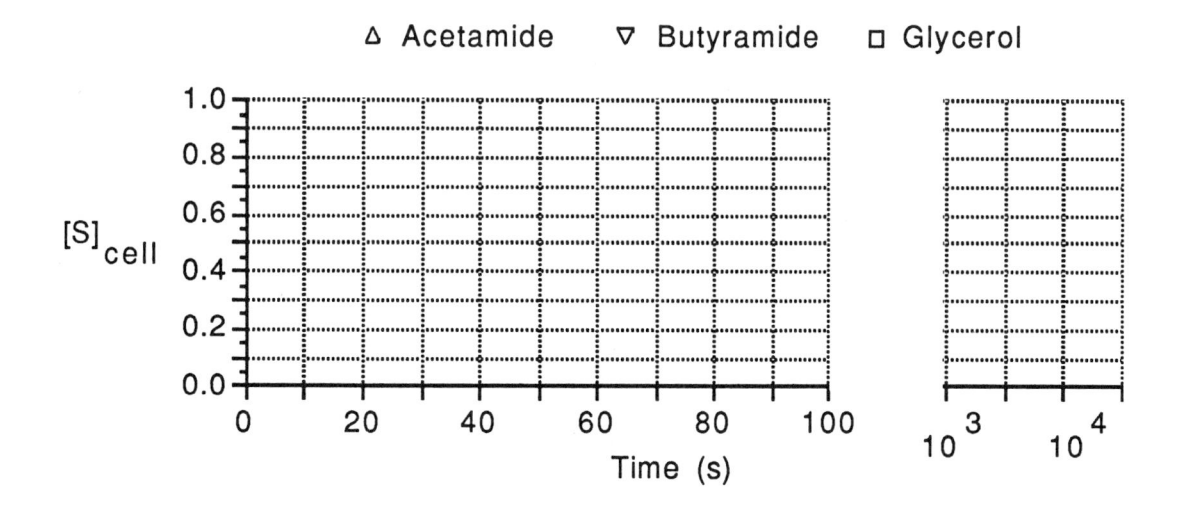

Chapter 6   Experiment 6.1

2. Is it possible to increase the diffusion rate of a molecule like glycerol by increasing the cell membrane area to cell volume ratio? Complete and graph the tables below (notice the geometric constraint on minimum surface area with respect to volume — cell B7).

   a. Assume that human RBCs are incubated in a $1M$ solution of glycerol. Complete the table below and then plot the intracellular concentration of glycerol at 2000 seconds as a function of cell area if cell volume is held constant at 9.7×10^{-11} cm^3.

V_{cell}	Area	$[S]_{cell}$ $t = 200$ s	$[S]_{cell}$ $t = 2000$ s
9.7×10^{-11}	1×10^{-6}	_____	_____
9.7×10^{-11}	2×10^{-6}	_____	_____
9.7×10^{-11}	3×10^{-6}	_____	_____
9.7×10^{-11}	4×10^{-6}	_____	_____
9.7×10^{-11}	5×10^{-6}	_____	_____
9.7×10^{-11}	6×10^{-6}	_____	_____
9.7×10^{-11}	7×10^{-6}	_____	_____
9.7×10^{-11}	8×10^{-6}	_____	_____
9.7×10^{-11}	9×10^{-6}	_____	_____

[S]cell vs. Surface Area x 10^6, $\Delta t = 2000$ s

   b. Assume that human RBCs are incubated in a $1M$ solution of glycerol. Complete the following table and then plot the intracellular concentration of glycerol at 2000 seconds as a function of cell volume if cell area is held constant at 1.42×10^{-6}

cm². $V_{cell} = 1.59 \times 10^{-10}$ cm³ corresponds to the volume of a sphere of area 1.42×10^2 cm², the upper limit for V_{cell}.

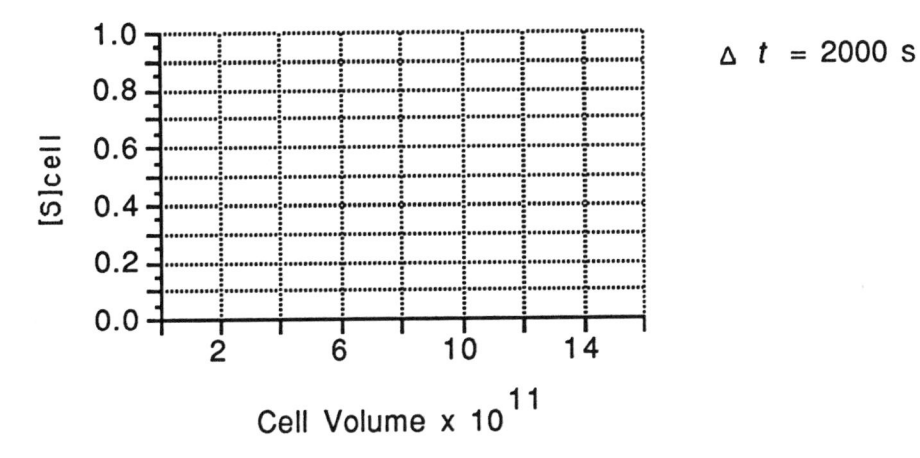

V_{cell}	Area	[S]$_{cell}$ $t = 200$ s	$t = 2000$ s
1.59×10^{-10}	1.42×10^{-6}	_____	_____
9.70×10^{-11}	1.42×10^{-6}	_____	_____
9.00×10^{-11}	1.42×10^{-6}	_____	_____
8.00×10^{-11}	1.42×10^{-6}	_____	_____
7.00×10^{-11}	1.42×10^{-6}	_____	_____
6.00×10^{-11}	1.42×10^{-6}	_____	_____
5.00×10^{-11}	1.42×10^{-6}	_____	_____
4.00×10^{-11}	1.42×10^{-6}	_____	_____
3.00×10^{-11}	1.42×10^{-6}	_____	_____
2.00×10^{-11}	1.42×10^{-6}	_____	_____
1.00×10^{-11}	1.42×10^{-6}	_____	_____

[Plot: [S]$_{cell}$ vs. Cell Volume × 10^{11}, Δ $t = 2000$ s]

c. What are the likely shapes for RBCs with large area/volume ratios?

3. Imagine a perfectly spherical cell. Determine the intracellular concentration of glycerol at 200 and 2000 seconds as a function of volume. Plot the values at 2000 seconds.

Because surface area is now a function of volume, the model described in Figure 6.3 should be modified so that the formula in cell B7 also exists in cell B3.

Chapter 6   Experiment 6.1

What is one disadvantage to being a large cell?

V_{cell}	Area	[S]$_{cell}$ $t = 200$ s	$t = 2000$ s
1 x 10^{-10}	_____	_____	_____
2 x 10^{-10}	_____	_____	_____
3 x 10^{-10}	_____	_____	_____
4 x 10^{-10}	_____	_____	_____
5 x 10^{-10}	_____	_____	_____
6 x 10^{-10}	_____	_____	_____
7 x 10^{-10}	_____	_____	_____
8 x 10^{-10}	_____	_____	_____
9 x 10^{-10}	_____	_____	_____

[Graph: [S]$_{cell}$ (y-axis, 0.0 to 1.0) vs. Cell Volume x 10^{10} (x-axis, 2 to 10); $\Delta t = 2000$ s]

Problems

1. Review your experience with the model described in Figure 6.3 and describe the limits that must be placed on the size and shape of cells if efficient transport of nutrients and wastes in and out of cells is to occur. You might want to consider (in a qualitative fashion) the metabolic costs of synthesizing membrane components as you formulate your answer.

    a. Why do you suppose living cells range in diameter from about 1 μm (bacteria) to about 100 μm (large animal cells)?

    b. Why are certain cells (e.g., mammalian RBCs) not spherical?

Passive Transport through Membranes

c. What metabolic features might contribute to the smaller size of bacterial cells as compared with animal cells?

2. A useful way to treat membrane-transport data is to consider half-times, the time required for the intracellular solute concentration to reach half of its final equilibrium level.

Because we presume that $[S]_{med}$ is essentially inexhaustible, $t = t_{1/2}$ (half-time) when:

$$[S]_{cell} = [S]_{med}/2 \tag{6.12}$$

Equation 6.12 may be substituted into Equation 6.10 to yield:

$$t_{1/2} = \frac{(\ln 2)(V_{cell})}{P_s A} \tag{6.13}$$

Modify the model described in Figure 6.3 so that it corresponds to the model described in Figure 6.4.

Figure 6.4

Formulas

	A	B
1	Ps =	1.E-7
2	Vcell =	9.70E-11
3	Area =	1.42E-6
4	[S]med =	1
5		
6		
7	Area (min) =	=(4*PI)*(((3*B2)/(4*PI))^(2/3))
8	t 1/2 =	=(LN(2)*B2)/(B1*B3)

(Continued.)

Chapter 6   Experiment 6.1

Figure 6.4 (Contd.)

Values

	A	B
1	Ps =	1.E-7
2	Vcell =	9.70E-11
3	Area =	1.42E-6
4	[S]med =	1
5		
6		
7	Area (min) =	1.02E-6
8	t 1/2 =	473.49

a. Calculate $t_{1/2}$ for acetamide, butyramide, and glycerol equilibra across the human RBC membrane. What is the relationship between $t_{1/2}$ and the permeability coefficient?

b. Using plots similar to those in Questions 2a and b, determine how $t_{1/2}$ changes with cell volume (constant membrane area). With membrane-area changes (constant cell volume). Plot your data for one of the compounds in a above.

c. Is there any information lost when $t_{1/2}$ alone is used to characterize a transport system of this type?

3. Substances that move across cell membranes are usually metabolized within the cell. This question permits you to explore this more complex situation.

Consider a single extracellular substrate that passively diffuses across a cell membrane where it is metabolized by a single-site, Michaelis enzyme (Figure 6.5).

Figure 6.5

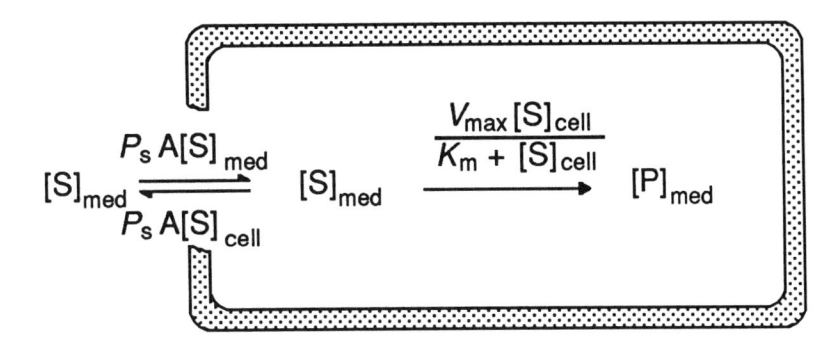

When passive diffusion of S across the membrane achieves steady-state equilibrium, $[S]_{cell}$ will be constant (Equation 6.14).

$$\frac{d[S]_{cell}}{dt} = \frac{P_s A [S]_{med} - P_s A [S]_{cell}}{V_{cell}} - \frac{V_{max} [S]_{cell}}{K_m + [S]_{cell}} = 0 \quad (6.14)$$

Equation 6.14 may be solved for $[S]_{cell}$ to yield:

$$[S]_{cell} = \frac{-b + \sqrt{b^2 - 4ac}}{2a} \quad (6.15)$$

Where $a = P_s A / V_{cell}$
$b = (P_s A K_m)/V_{cell} + V_{max} - ([S]_{med} P_s A)/V_{cell}$
$c = -(P_s A K_m [S]_{med})/V_{cell}$

Build and verify the model described in Figure 6.6 and then answer the following questions:

a. What happens to $[S]_{cell}$ as the concentration of enzyme within the cell increases? Represent enzyme concentration with a range of V_{max} values from 10^{-2} to 10^{-8} mol/lit-s. You might want to add a few additional points for V_{max} values between 10^{-3} and 10^{-4} mol/lit-s.

b. The flux of S to P per unit of cell volume (mol/lit-s) is an excellent measure of the flow of S into a cell and its ultimate

Chapter 6   Experiment 6.1

utilization. If $[S]_{med}$ is not limiting can flux/volume increase indefinitely as enzyme concentration increases?

c. How does K_m affect the results in a and b above?

d. How might a cell maintain a high metabolic flux and a low concentration of metabolite intermediates (e.g., $[S]_{cell}$) by regulating both K_m and V_{max}?

e. How do changes in cell area or volume influence metabolic flux?

Figure 6.6

Formulas

	A	B
1	Ps =	2.10E-8
2	Vcell =	9.70E-11
3	Area =	1.40E-6
4	[S]med =	1
5	Vmax =	1.00E-2
6	Km =	1.00E-3
7		=(B1*B3*B6/B2)+B5-(B1*B3*B4/B2)
8	[S]cell =	=(-B7+SQRT((B7*B7)+(4*B1*B3*B6*B4*B1*B3/(B2*B2))))/(2*B1*B3/B2)
9	Flux/Vol =	=(B5*B8)/(B8+B6)

Values

	A	B
1	Ps =	2.10E-8
2	Vcell =	9.70E-11
3	Area =	1.40E-6
4	[S]med =	1
5	Vmax =	1.00E-2
6	Km =	1.00E-3
7		9.70E-3
8	[S]cell =	3.13E-5
9	Flux/Vol =	3.03E-4

Experiment 6.2

Facilitated Diffusion through Membranes

Membrane systems contain carrier proteins that can bind solute and catalyze solute transfer across the membrane (Figure 6.7).

The flux through the membrane is proportional to the solute protein bound by the carrier. The mechanism may be written:

$$S_{med} + carrier \underset{k_2}{\overset{k_1}{\rightleftharpoons}} S_{med} \cdot carrier \overset{k_3}{\longrightarrow} S_{cell} + carrier$$

Considering the reverse reaction as well, the mechanism may be written:

$$S_{med} + carrier \underset{k_2}{\overset{k_1}{\rightleftharpoons}} S_{med} \cdot carrier \underset{k_4}{\overset{k_3}{\rightleftharpoons}} S_{cell} \cdot carrier \underset{k_6}{\overset{k_5}{\rightleftharpoons}} S_{cell} + carrier$$

Figure 6.7

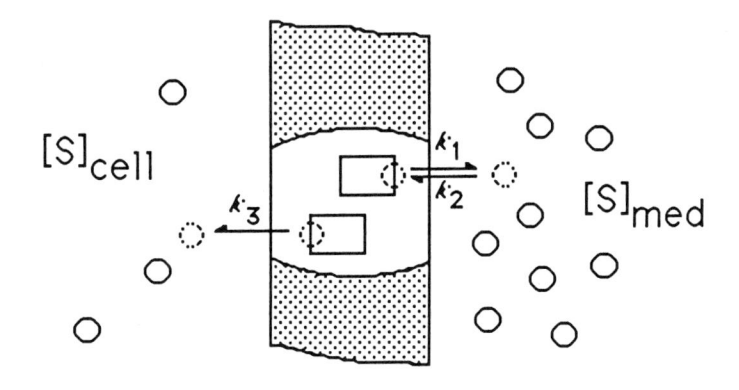

Chapter 6   Experiment 6.2

Or, if the carrier is symmetrical (and $k_1 = k_6$, $k_2 = k_5$, $k_3 = k_4$):

$$S_{med} + carrier \underset{k_2}{\overset{k_1}{\rightleftharpoons}} S_{med} \cdot carrier \underset{k_3}{\overset{k_3}{\rightleftharpoons}} S_{cell} \cdot carrier \underset{k_1}{\overset{k_2}{\rightleftharpoons}} S_{cell} + carrier$$

If you assume that the carrier is symmetrical and that association and dissociation of ligand (k_1 and k_2) is rapid with respect to transport (k_3), then the Michaelis treatment developed in Chapter 4 becomes the foundation for a treatment of facilitated diffusion. In Chapter 4, however, only initial reaction rates (i.e., reaction rates near time = 0) were considered. In this chapter the Michaelis treatment is expanded to include all points in time, and net solute transport rates with time is our principal experimental probe.

Consider the net flux of substrate from the medium into the cell J_{net}:

$$\begin{aligned} J_{net} &= J_{med \to cell} - J_{cell \to med} \\ &= k_3 ([S_{med} \cdot carrier] - [S_{cell} \cdot carrier]) \end{aligned} \quad (6.16)$$

Because $k_2/k_1 = K_m$ under the assumptions of the Michaelis treatment

$$\frac{k_2}{k_1} = K_m = \frac{[S]_{med}[carrier]}{[S_{med} \cdot carrier]} = \frac{[S]_{cell}[carrier]}{[S_{cell} \cdot carrier]} \quad (6.17)$$

You can rewrite Equation 6.16 as

$$J_{net} = k_3 \left(\frac{[S]_{med}[carrier]}{K_m} - \frac{[S]_{cell}[carrier]}{K_m} \right) \quad (6.18)$$

All that remains is to find an expression for [carrier] in terms of the measurable quantities $[S]_{med}$, $[S]_{cell}$, K_m, and $[carrier]_{tot}$. Because mass must be conserved, total carrier ($[carrier]_{tot}$) may be expressed as:

$$[carrier]_{tot} = [S_{med} \cdot carrier] + [S_{cell} \cdot carrier] + [carrier] \quad (6.19)$$

If Equation 6.17 is used to obtain expressions for [S_{med} · carrier] and [S_{cell} · carrier], Equation 6.19 can be solved for [carrier] to yield:

$$[\text{carrier}] = \frac{K_m [\text{carrier}]_{tot}}{(K_m + [S]_{cell} + [S]_{med})} \qquad (6.20)$$

The expression for [carrier] in Equation 6.20 can be substituted into Equation 6.18 to yield:

$$J_{net} = \frac{d[S]_{cell}}{dt}$$

$$= k_3 [\text{carrier}]_{tot} \left(\frac{[S]_{med}}{K_m + [S]_{cell} + [S]_{med}} - \frac{[S]_{cell}}{K_m + [S]_{cell} + [S]_{med}} \right) \qquad (6.21)$$

$d[S]_{cell}/dt$ is substituted for J_{net} to facilitate a small amount of calculus: Equation 6.22 is integrated to yield Equation 6.23 (where [S]$_{med}$ is assumed to occupy a large volume and is therefore a constant).

$$\frac{d[S]_{cell}}{\left(\frac{[S]_{med} - [S]_{cell}}{K_m + [S]_{cell} + [S]_{med}} \right)} = k_3 [\text{carrier}]_{tot} dt \qquad (6.22)$$

$$[S]_{cell} + (2 [S]_{med} + K_m) \left[\ln \left(1 - \frac{[S]_{cell}}{[S]_{med}} \right) \right] = - k_3 [\text{carrier}]_{tot} t \qquad (6.23)$$

Unlike Equation 6.10, Equation 6.23 cannot be solved explicitly for [S]$_{cell}$ — numerical methods such as Newton's method are required. However, $t_{1/2}$ may be calculated by substituting the relationship [S]$_{cell}$ = [S]$_{med}$/2 (Equation 6.12) into Equation 6.23, giving

$$t_{1/2} = \frac{(2 [S]_{med} + K_m) \ln 2 - [S]_{med}/2}{k_3 [\text{carrier}]_{tot}} \qquad (6.24)$$

Equation 6.24 can be used to describe the transport of the amino acid methionine in human RBCs. Transport systems for several other amino

Chapter 6   Experiment 6.2

acids, sugars, and metabolites in various cell types can also be represented by Equation 6.24.

Directions

Build and verify the model described in Figure 6.8.

Questions

1. How does the half-time of equilibration for methionine change with the concentration of methionine in the medium?

   Plot the $t_{1/2}$ as a function of $[S]_{med}$ for methionine.

   (The model described in Figure 6.8 already contains Km, k3, and [Carrier]tot values appropriate for methionine transport in human RBCs. Concentrations are given in molar terms, the rate constant k3 in min-1, and half-times in minutes.)

Figure 6.8

Formulas

	A	B
1	Km =	5.2E-3
2	K3 =	5.6E2
3	[Carrier]tot =	1.0E-6
4	[S]med =	2.0E-5
5		
6	t 1/2 =	=((((2*B4)+B1)*LN(2))-(B4/2))/(B2*B3)

Values

	A	B
1	Km =	5.2E-3
2	K3 =	5.6E2
3	[Carrier]tot =	1.0E-6
4	[S]med =	2.0E-5
5		
6	t 1/2 =	6.4680

$[S]_{med}$	log $[S]_{med}$	$t_{1/2}$
1.0×10^{-6}	-6	_____
1.0×10^{-5}	-5	_____
1.0×10^{-4}	-4	_____
1.0×10^{-3}	-3	_____
1.0×10^{-2}	-2	_____
1.2×10^{-2}	-1.92	_____
1.5×10^{-2}	-1.82	_____
1.7×10^{-2}	-1.77	_____
2.0×10^{-2}	-1.70	_____

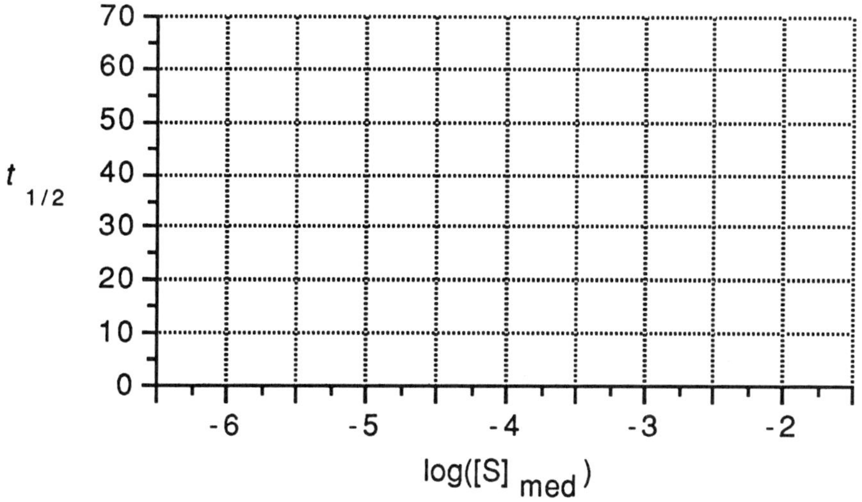

Why does $t_{1/2}$ approach a limiting value as $[S]_{med}$ decreases?

2. How does the half-time of equilibration for methionine change with K_m of the carrier? Complete and graph the table below.

		$t_{1/2}$	
K_m	$[S]_{med}$ = 1 m*M*	$[S]_{med}$ = 5 m*M*	$[S]_{med}$ = 10 m*M*
0.001	_____	_____	_____
0.002	_____	_____	_____
0.003	_____	_____	_____
0.004	_____	_____	_____
0.005	_____	_____	_____
0.006	_____	_____	_____
0.007	_____	_____	_____
0.008	_____	_____	_____
0.009	_____	_____	_____

Chapter 6   Experiment 6.2

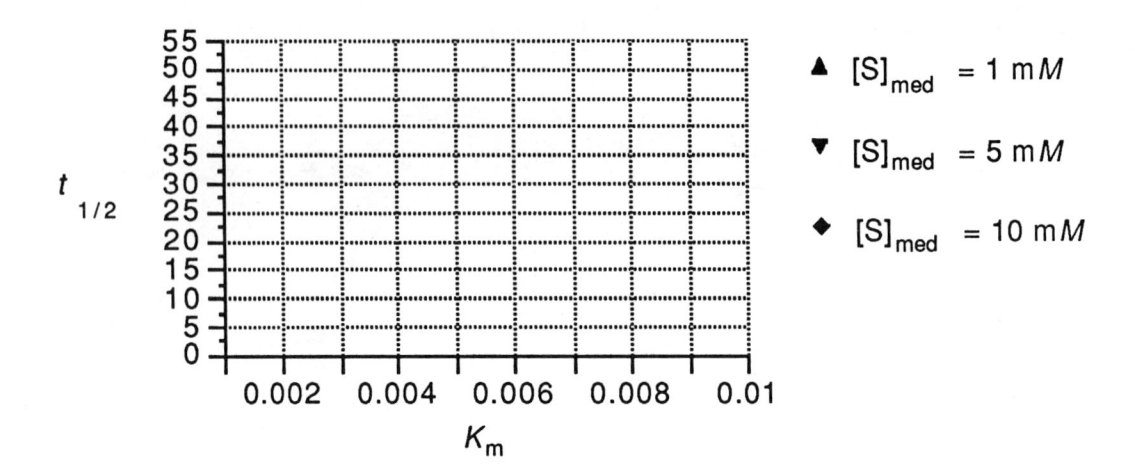

[Graph: $t_{1/2}$ vs K_m, with legend:
▲ $[S]_{med} = 1\ mM$
▼ $[S]_{med} = 5\ mM$
◆ $[S]_{med} = 10\ mM$]

Is transport optimized by high or low values of K_m in the preceding table?

3. After reflecting on the results, can you suggest ways to improve on nature and speed the rate of methionine transport into RBCs by either altering the K_m of the carrier or the steady-state level of methionine in the extracellular medium?

Problems

1. The amino acid methionine is used in RBCs to form S-adenosyl-methionine. S-adenosylmethionine is an important methyl-donor in protein methylation reactions. The net conversions in this reaction are protein to protein-CH_3 and plasma methionine to plasma homocysteine (Figure 6.9).

Facilitated Diffusion through Membranes

Figure 6.9

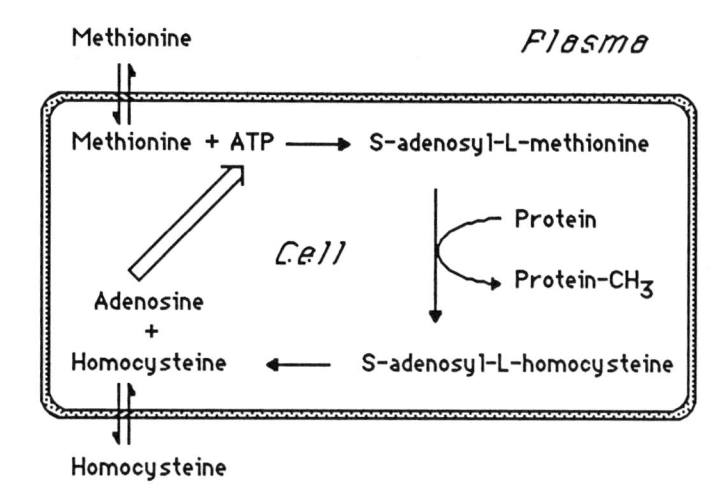

```
           Methionine                        Plasma
               ↓↑
    ┌──────────────────────────────────────────────┐
    │  Methionine + ATP ────► S-adenosyl-L-methionine │
    │                                               │
    │              ↗                   ┌─ Protein   │
    │            /         Cell        │            │
    │          /                       └─► Protein-CH₃ │
    │  Adenosine                                    │
    │     +                                         │
    │  Homocysteine ◄──── S-adenosyl-L-homocysteine │
    │     ↓↑                                        │
    └──────────────────────────────────────────────┘
           Homocysteine
```

The steady-state rate of these reactions has been estimated to be about 1000 molecules/cell/minute. Plasma methionine concentration is 20 µM. The methionine carrier has a $K_m = 5.2$ mM for methionine. Assume that k_3 is 560 min^{-1}.

Build a model that calculates the number of methionine transporters required to maintain a specific cellular concentration of methionine. How many methionine transporters are required if the cellular concentration of methionine must be maintained at 18 µM? 10 µM? 2 µM?

Assume that the molecular weight of the carrier is 50,000 and that RBC membranes contain about 4 x 10^{-13} g protein/cell. What fraction of the membrane protein must be methionine carrier for each case?

How would a change in K_m of the carrier affect these results? In k_3?

Experiment 6.3

Passive versus Facilitated Transport

Passive and facilitated (carrier-mediated) transport processes exhibit very different behaviors. In this experiment you will have the opportunity to explore these differences.

The rate equations for passive and facilitated transport were developed in Experiments 6.1 and 6.2.

Passive transport (from Equation 6.7):

$$\frac{d[S]_{cell}}{dt} = \frac{P_s A}{V_{cell}}([S]_{med} - [S]_{cell}) \tag{6.25}$$

Facilitated transport (from Equation 6.21):

$$\frac{d[S]_{cell}}{dt} = k_3 [carrier]_{tot} \left(\frac{[S]_{med}}{K_m + [S]_{med}} - \frac{[S]_{cell}}{K_m + [S]_{cell}} \right) \tag{6.26}$$

It is convenient to consider facilitated transport in terms of the density of the carriers in molecules/cm^2 that mediate transport. Because

$$[carrier]_{tot} = \frac{density \cdot 1000 \cdot A}{6.02 \times 10^{23} \cdot V_{cell}} \tag{6.27}$$

Equation 6.26 may be rewritten:

$$\frac{d[S]_{cell}}{dt} = \frac{k_3 A}{V_{cell}} \left(\frac{density}{6.02 \times 10^{20}} \right) \left(\frac{[S]_{med}}{K_m + [S]_{med}} - \frac{[S]_{cell}}{K_m + [S]_{cell}} \right) \tag{6.28}$$

The equations for facilitated transport do not yield exact solutions for $[S]_{cell}$ (see derivation of Equation 6.24 in Experiment 6.2); therefore in this experiment you will compare passive with facilitated transport rates under

Passive versus Facilitated Transport

initial conditions when $[S]_{cell}$ is close to zero and Equations 6.25 and 6.28 can be approximated by Equations 6.29 and 6.30, respectively.

$$\frac{d[S]_{cell}}{dt} = \frac{P_s A}{V_{cell}}([S]_{med}) \tag{6.29}$$

$$\frac{d[S]_{cell}}{dt} = \frac{k_3 A}{V_{cell}}\left(\frac{density}{6.02 \times 10^{20}}\right)\left(\frac{[S]_{med}}{[S]_{med} + K_m}\right) \tag{6.30}$$

Directions

1. Build and verify the model described in Figure 6.10.

2. Note that the model includes cells to calculate conditions in which both passive and facilitated transport processes are functioning.

Figure 6.10

Formulas

	A	B	C	D	E
1		Area =			1.42E-6
2		Vcell =			9.70E-11
3		Ps =			2.10E-8
4		k3 =			560
5		Carrier Density =			1.00E9
6		Km =			5.00E-3
7		[S]med =			1.00E-4
8					
9		Initial values of:			
10		Passive d[S]cell/dt =	=B1*B3*B7/B2		
11		Facilitated d[S]cell/dt =	=(B4*B5*B1/(B2*6.02E20))*(B7/(B7+B6))		
12		Total d[S]cell/dt =	=B10+B11		

(Continued.)

Chapter 6   Experiment 6.3

Figure 6.10 (Contd.)

Values

	A	B	C	D	E
1		Area =			1.42E-6
2		Vcell =			9.70E-11
3		Ps =			2.10E-8
4		k3 =			560
5		Carrier Density =			1.00E9
6		Km =			5.00E-3
7		[S]med =			1.00E-4
8					
9		Initial values of:			
10		Passive d[S]cell/dt =			3.07E-8
11		Facilitated d[S]cell/dt =			2.67E-7
12		Total d[S]cell/dt =			2.98E-7

Questions

Explore passive, facilitated, and combined transport rates for various initial concentrations of $[S]_{med}$. Complete the table below for glycerol transport across a human RBC membrane and use the kinetic parameters identified in Figure 6.10.

$[S]_{med}$	Passive	Facilitated	Combined
0.01	_____	_____	_____
0.02	_____	_____	_____
0.03	_____	_____	_____
0.04	_____	_____	_____
0.05	_____	_____	_____
0.06	_____	_____	_____
0.07	_____	_____	_____
0.08	_____	_____	_____
0.09	_____	_____	_____
0.10	_____	_____	_____

Passive versus Facilitated Transport

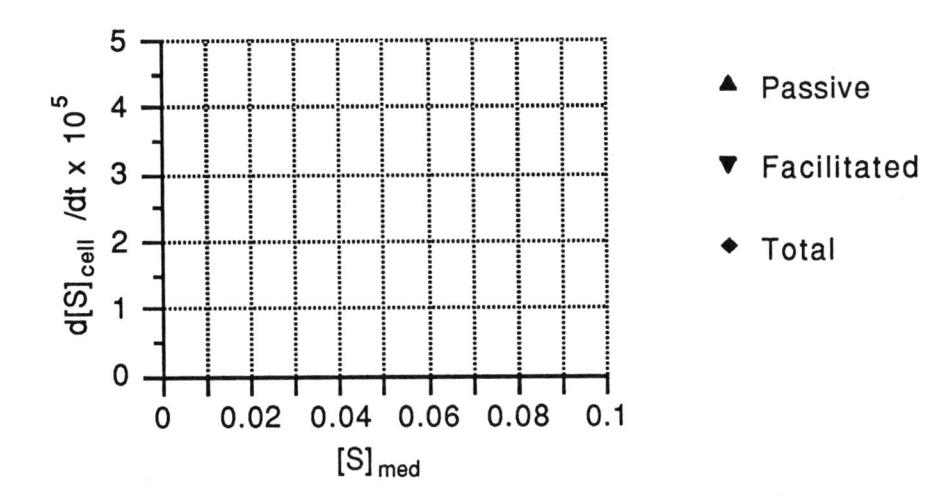

Why does the initial rate of carrier-mediated transport level out at high values of [S]$_{med}$?

Problems

1. Suppose that you measured the initial rate of ethanol transport in human erythrocytes and obtained the following data:

[Ethanol]$_{med}$ μM)	Initial Transport Rate (μmol/l cells ≥ min)
0	0
1	1.15
2	2.42
5	5.98
10	12.05
20	24.00

   What is the mechanism of transport? Compute the relevant K_m, P_s, and/or k_3 values.

2. If the K_m of a transport carrier is large (say five-fold over the largest possible value of [S]$_{med}$), would it still be possible to answer Question 1?

Chapter 6 Experiment 6.3

3. Given the following data, what can you conclude about transport mechanism(s)? Find the values of K_m, P_s, K_3, and carrier density if applicable and possible. Assume values of area and V_{cell} as in Figure 6.10.

$[S]_{med}$	Initial Transport Rate (x 10^7)
0.0001	0.000
0.001	0.055
0.002	0.108
0.005	0.258
0.01	0.479
0.02	0.841
0.05	1.582
0.1	2.350
0.2	3.410
0.5	5.870
1.0	9.630

Experiment 6.4*

Active Transport across Membranes: Calculation of ΔG'

Experiments 6.1 and 6.2 describe transport systems in which at equilibrium $[S]_{cell} = [S]_{med}$. Solute is not concentrated either within or without the cell.

Let's consider a transport system that actively concentrates solute on one side of the membrane and, as a consequence, creates a free energy difference across the membrane.

$[S]_{cell}$ and $[S]_{med}$ contribute to the free-energy difference across a membrane according to the relationship:

$$\Delta G' = \Delta G^{\circ\prime} + RT \ln \frac{[S]_{cell}}{[S]_{med}} \tag{6.31}$$

where $\Delta G^{\circ\prime}$ = the standard free energy change for the reaction $S_{med} \to S_{cell}$

R = the gas constant (1.98 cal/°K mol)

T = temperature (°K)

Because the chemical potential of the solute at standard conditions (1M) is essentially the same in the outside medium as in the intracellular environment (Experiment 6.6 provides one exception), $\Delta G^{\circ\prime} = 0$.

Thus

$$\Delta G' = RT \ln \frac{[S]_{cell}}{[S]_{med}} \tag{6.32}$$

Chapter 6   Experiment 6.4*

Figure 6.11

Formulas

	A	B
1	[S]cell =	100
2	[S]med =	1
3	R =	0.00198
4	T =	298
5	ΔGo' =	0
6		
7	ΔG' =	=B5+(B3*B4*LN(B1/B2))

Values

	A	B
1	[S]cell =	100
2	[S]med =	1
3	R =	0.00198
4	T =	298
5	ΔGo' =	0
6		
7	ΔG' =	2.7172

Directions

1. Build and verify the model described in Figure 6.11.

2. The model is based on Equation 6.31 with the $\Delta G^{o\prime}$ term included. In most instances, however, $\Delta G^{o\prime}$ (cell B4) equals zero.

Questions

1. Use the model described in Figure 6.11 to create a semilog plot of $[S]_{cell}/[S]_{med}$ vs. $\Delta G'$.

$[S]_{cell}$	$[S]_{med}$	$[S]_{cell}/[S]_{med}$	$\Delta G'$
0.001	1	___	___
0.01	1	___	___
0.1	1	___	___
1	1	___	___
10	1	___	___
100	1	___	___
1000	1	___	___

Calculation of ΔG′

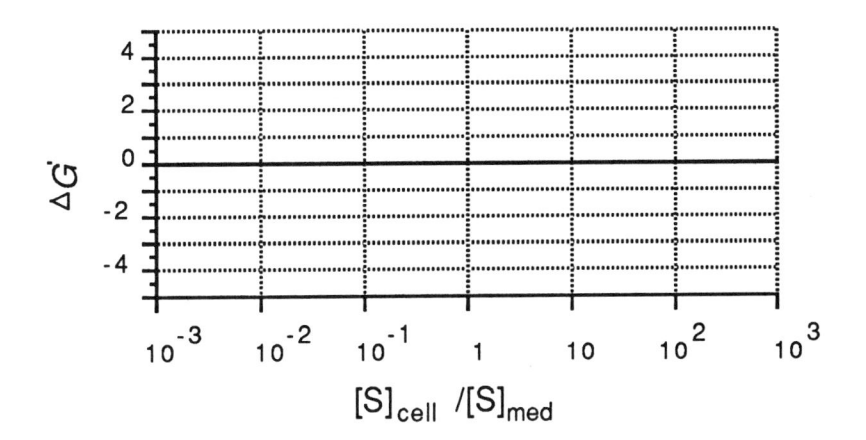

[Graph: ΔG vs $[S]_{cell}/[S]_{med}$, x-axis logarithmic from 10^{-3} to 10^{3}, y-axis from -4 to 4]

2. What does it cost the cell in energy to concentrate a solute 10-fold? 100-fold? 1000-fold?

3. What is the meaning of positive and negative values for free energy?

*Experiment 6.5**

Active Transport across Membranes: Distribution of Solute

Active transport systems can be driven by coupling either chemical or electrical energy to the transport reaction. Consider a reaction in which the transport of S into the cell is linked to the hydrolysis of ATP (Figure 6.12).

The free energy of this reaction may be written:

$$\Delta G' = \Delta G^{\circ'}_1 + \Delta G^{\circ'}_2 + RT \ln \frac{[ADP][P_i]}{[ATP]} + RT \ln \frac{[S]_{cell}}{[S]_{med}} \tag{6.33}$$

where

$\Delta G^{\circ'}_1$ = standard free energy contribution of the chemical potential of the solute

= 0 kcal/mol

$\Delta G^{\circ'}_2$ = standard free energy contribution of the hydrolysis of ATP

= -7.3 kcal/mol

This leads to

$$\Delta G' = 0 - 7.3 + RT \ln \frac{[ADP][P_i]}{[ATP]} + RT \ln \frac{[S]_{cell}}{[S]_{med}} \tag{6.34}$$

Figure 6.12

$$H_2O + ATP_{cell} + S_{med} \rightarrow ADP_{cell} + P_{i\ cell} + S_{cell}$$

Distribution of Solute

If we assume that ATP, ADP, and P_i are maintained at standard state concentrations ($1M$), that is,

$$RT \ln \frac{[ADP][P_i]}{[ATP]} = 0 \qquad (6.35)$$

Then

$$\Delta G' = \Delta G_2^{\circ'} + RT \ln \frac{[S]_{cell}}{[S]_{med}} = -7.3 + RT \ln \frac{[S]_{cell}}{[S]_{med}} \qquad (6.36)$$

At 25°C then, the ratio of cell to medium solute at equilibrium ($\Delta G' = 0$) would be:

$$\frac{[S]_{cell}}{[S]_{med}} = e^{\left(\frac{-\Delta G_2^{\circ'}}{0.00198 \times 298}\right)} = e^{\left(\frac{7.3}{0.00198 \times 298}\right)} \qquad (6.37)$$

Directions

Build and verify the model described in Figure 6.13.

Questions

1. Explore the model as initialized in Figure 6.13. What fold-concentration can be accomplished by linking the transport reaction to a single ATP hydrolysis event? Does this amaze you?

Figure 6.13

Formulas

	A	B
1	ΔG°2 =	-7.3
2	R =	0.00198
3	T =	298
4		
5	[S]cell/[S]med =	=EXP(-B1/(B2*B3))

Values

	A	B
1	ΔG°2 =	-7.3
2	R =	0.00198
3	T =	298
4		
5	[S]cell/[S]med =	236108

Chapter 6   Experiment 6.5*

2. What would be the equilibrium distribution of solute if the transport reaction shown in Figure 6.12 were coupled to the hydrolysis of 3 moles of ATP rather than 1 mole?

3. What would be the equilibrium distribution of solute if the stoichiometry of the transport reaction were 3 moles of solute transported per 1 mole of ATP?

4. How would the results obtained in Questions 1 and 2 differ if ATP were replaced by phosphoenolpyruvate ($\Delta G^{o'}$ = - 14.8 kcal/mol) or glycerol-1-phosphate ($\Delta G^{o'}$ = - 2.2 kcal/mol).

(Note: In cells, ATP is frequently used as an energy source for transport reactions. However, the active transport of several sugars in certain bacteria is coupled to phosphoenolpyruvate hydrolysis).

Problems

In living cells, ATP, ADP, and inorganic PO_4 are not maintained at $1M$ concentrations. Typically, the ratio of [ATP]/[ADP] is fixed at about 7, and the concentration of PO_4 is about 10^{-3} M. Use the model described in Figure 6.14 to calculate the equilibrium distribution of solute across the membrane for ATP-driven reactions under these physiologic conditions. How would a change in the [ATP]/[ADP] ratio or [PO_4] affect the ratio of [S]$_{cell}$/[S]$_{med}$ at equilibrium?

Distribution of Solute

Figure 6.14

Formulas

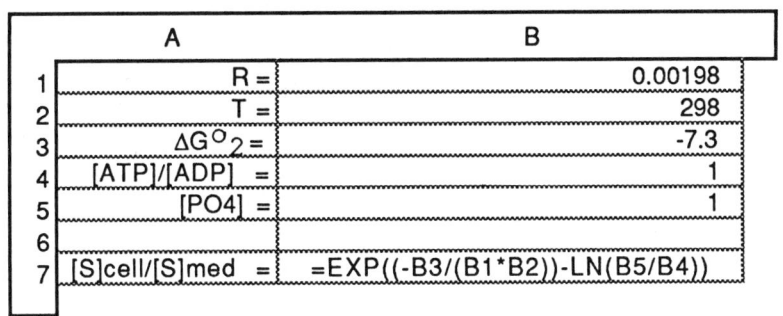

	A	B
1	R =	0.00198
2	T =	298
3	ΔG°_2 =	-7.3
4	[ATP]/[ADP] =	1
5	[PO4] =	1
6		
7	[S]cell/[S]med =	=EXP((-B3/(B1*B2))-LN(B5/B4))

Values

	A	B
1	R =	0.00198
2	T =	298
3	ΔG°_2 =	-7.3
4	[ATP]/[ADP] =	1
5	[PO4] =	1
6		
7	[S]cell/[S]med =	236108

Experiment 6.6*

Active Transport across Membranes: Distribution of an Ionic Solute in the Presence of an Electrical Field

Consider the transport of an ionic species (e.g., Na+, HCO$_3^-$) across a membrane bilayer in the presence of an electrical field. The energetics of this reaction is described by Equation 6.38:

$$\Delta G' = \Delta G^{\circ'} + RT \ln \frac{[S]_{cell}}{[S]_{med}} + z\, F\Delta\Psi_{cell\text{-}med} \qquad (6.38)$$

where
- z = charge on translocated molecule
- F = Faraday (23 kcal/volt)
- $\Delta\Psi_{cell\text{-}med}$ = difference in electrical potential across the membrane

At equilibrium ($\Delta G' = 0$)

$$\frac{[S]_{cell}}{[S]_{med}} = e^{-\left(\frac{\Delta G^{\circ'} + z\, F\Delta\Psi_{cell\text{-}medium}}{RT}\right)} \qquad (6.39)$$

Notice that a voltage applied across the membrane $\Delta\Psi_{cell\text{-}med}$ affects the distribution of the charged solute. A gradient of both solute and electrical charge across a membrane is called an *electrochemical gradient*. In this experiment you will build and explore models of electrochemical gradients.

In many instances $\Delta G^{\circ'}$ is zero. However, if ion transport is coupled to another reaction (e.g., ATP hydrolysis) additional terms must be considered. For example, the reaction

$$H_2O + ATP + [S]_{med} \rightarrow ADP + P_i + [S]_{cell}$$

would be written:

$$\Delta G' = \Delta G^{\circ\prime} + RT \ln \frac{[S]_{cell}}{[S]_{med}} + RT \ln \frac{[ADP][P_i]}{[ADP]} + zF\Delta\Psi_{cell-med} \quad (6.40)$$

where $\Delta G^{\circ\prime} = -7.3$ Kcal/mol (as usual)

At equilibrium, $\Delta G' = 0$, and Equation 6.40 can be written:

$$RT \ln \frac{[S]_{cell}}{[S]_{med}} = -\Delta G^{\circ\prime} - zF\Delta\Psi_{cell-med} - RT \ln \frac{[ADP][P_i]}{[ADP]} \quad (6.41)$$

or

$$\frac{[S]_{cell}}{[S]_{med}} = e^{-\left(\frac{\Delta G^{\circ\prime} - zF\Delta\Psi_{cell-med} + RT \ln \frac{[ADP][P_i]}{[ADP]}}{RT}\right)} \quad (6.42)$$

Directions

1. Build and verify the model described in Figure 6.15. This model represents the situation described by Equation 6.39.

2. Model values are initialized for a potential gradient, $\Delta\Psi_{cell-med}$, of -0.1 volts — a value that is typical of many cell types.

Chapter 6   Experiment 6.6*

Figure 6.15

Formulas

	A	B
1	R =	0.00198
2	T =	298
3	F =	23
4	ΔG(0') =	-7.3
5	Δ Psi =	-0.1
6	Charge =	1
7		
8	[S]cell/[S]med =	=EXP((-B4-(B5*B6*B3))/(B1*B2))

Values

	A	B
1	R =	0.00198
2	T =	298
3	F =	23
4	ΔG(0') =	-7.3
5	Δ Psi =	-0.1
6	Charge =	1
7		
8	[S]cell/[S]med =	1.16E7

Questions

Use the model described in Figure 6.15 to determine the effect of the voltage across the membrane on the ratio $[S]_{cell}/[S]_{med}$ when:

a. The migrating species is a monovalent cation (e.g., Na^+).

$\Delta G^{o'}$	Charge	$\Delta \psi_{cell-med}$	$[S]_{cell}/[S]_{med}$
0	1	-0.2	_____
0	1	-0.16	_____
0	1	-0.12	_____
0	1	-0.08	_____
0	1	-0.04	_____
0	1	0	_____
0	1	0.04	_____
0	1	0.08	_____
0	1	0.12	_____
0	1	0.16	_____
0	1	0.2	_____

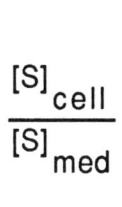

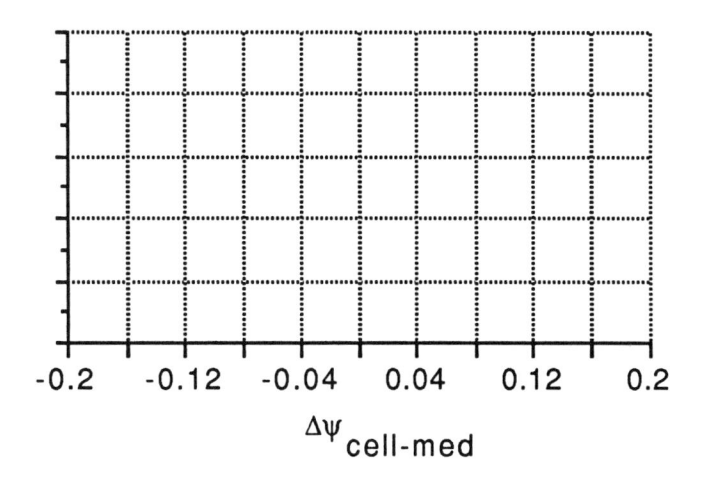

Distribution of an Ionic Solute

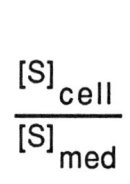

[S]$_{cell}$/[S]$_{med}$ vs $\Delta\psi_{cell-med}$ (graph with x-axis from -0.2 to 0.2)

b. The migrating species is a divalent cation (e.g., Ca++).

$\Delta G^{o\prime}$	Charge	$\Delta\psi_{cell-med}$	[S]$_{cell}$/[S]$_{med}$
0	2	-0.2	_____
0	2	-0.16	_____
0	2	-0.12	_____
0	2	-0.08	_____
0	2	-0.04	_____
0	2	0	_____
0	2	0.04	_____
0	2	0.08	_____
0	2	0.12	_____
0	2	0.16	_____
0	2	0.2	_____

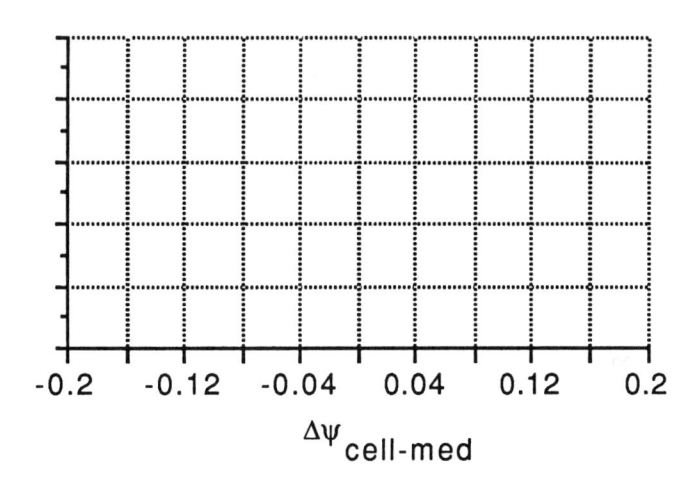

[S]$_{cell}$/[S]$_{med}$ vs $\Delta\psi_{cell-med}$ (graph with x-axis from -0.2 to 0.2)

Chapter 6   Experiment 6.6*

c. The migrating species is a monovalent anion (e.g., HCO_3^-).

$\Delta G^{o'}$	Charge	$\Delta\psi_{cell-med}$	$[S]_{cell}/[S]_{med}$
0	___	-0.2	_____
0	___	-0.16	_____
0	___	-0.12	_____
0	___	-0.08	_____
0	___	-0.04	_____
0	___	0	_____
0	___	0.04	_____
0	___	0.08	_____
0	___	0.12	_____
0	___	0.16	_____
0	___	0.2	_____

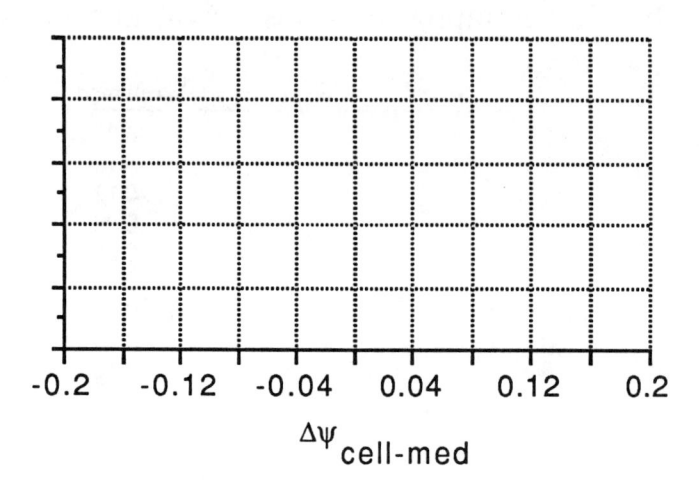

$\dfrac{[S]_{cell}}{[S]_{med}}$

$\Delta\psi_{cell-med}$: -0.2, -0.12, -0.04, 0.04, 0.12, 0.2

d. The migrating species is a zwitterion (e.g., glycine: NH_3^+-CH_2-COO^-).

$\Delta G^{o'}$	Charge	$\Delta\psi_{cell-med}$	$[S]_{cell}/[S]_{med}$
0	___	-0.2	_____
0	___	-0.16	_____
0	___	-0.12	_____
0	___	-0.08	_____
0	___	-0.04	_____
0	___	0	_____
0	___	0.04	_____
0	___	0.08	_____
0	___	0.12	_____
0	___	0.16	_____
0	___	0.2	_____

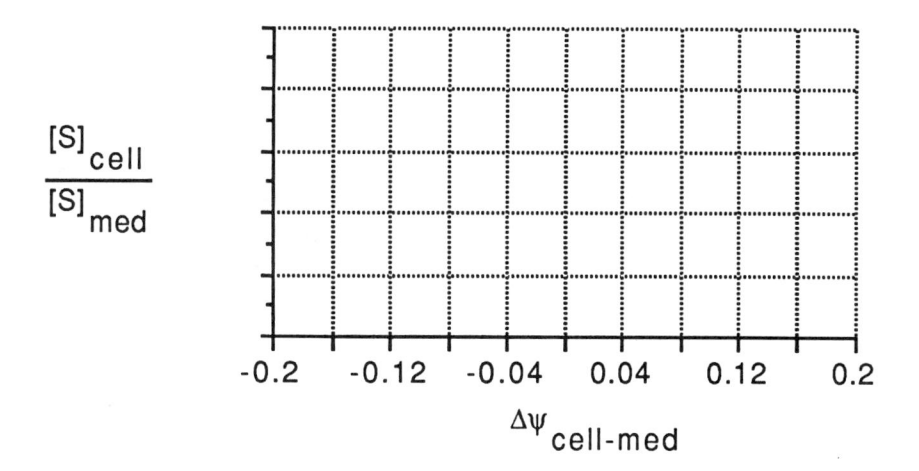

Problems

1. Under physiologic conditions, the ratio of [ATP]/[ADP] is about 7 and the concentration of phosphate is about 1 mM (see Experiment 6.5). Build a model that gives the ratio of an ion in the cellular and extracellular medium when the concentration of ATP, ADP, and P$_i$ is *not* at standard state.

2. In some transport reactions, more than one ion can be translocated and the stoichiometry is not 1:1. For instance, a very important reaction in living cells is catalyzed by sodium-potassium ATPase:

$$3Na^+_{cell} + 2K^+_{med} + ATP \rightarrow 3Na^+_{med} + 2K^+_{cell} + ADP + P_i$$

Build a model for this reaction that describes the distribution of sodium and potassium ions under physiologic conditions.

The energetic equation for the reaction:

$$aA_{med} + bB_{med} \rightarrow aA_{cell} + bB_{med}$$

is

$$\Delta G' = \Delta G^{o\prime} + RT \ln \frac{[A]^a_{cell}[B]^b_{med}}{[A]^a_{med}[B]^b_{cell}} + az_A F \Delta \Psi_{cell-med} + bz_B F \Delta \Psi_{cell-med}$$

Chapter 7

Structure and Stability of Macromolecules

Macromolecules require specific three-dimensional conformations to function properly. These conformations result primarily from intra-molecular interactions between atoms in a macromolecule and intermolecular interactions between the macromolecule and the surrounding solvent. Although the folding process appears to be quite complex, the instructions guiding this process are generally specified entirely by the primary sequence of the protein or nucleic acid. Many denatured proteins and nucleic acids, for example, spontaneously refold into functional conformations once denaturing conditions are removed.

In this chapter you will explore some of the noncovalent interactions that are responsible for maintaining specific three-dimensional conformations in macromolecules. You will build and interrogate models of steric hindrance (van der Waals interactions), electrostatic forces, and the hydrophobic effect.

The environment within which three-dimensional structures are maintained is complex, and the models you will build are only approximations of true behavior. In many instances, and especially with respect to the hydrophobic effect, quantitative treatments are very preliminary.

Nevertheless, you can use even approximate treatments to acquire an appreciation of the forces that help define a three-dimensional structure.*

---

* Unless otherwise specified, energy units will be expressed in kcal/mol throughout this chapter, and distances will be expressed in Angstroms (Å).

Experiment 7.1

Three-Dimensional Structures of Macromolecules:

Restrictions on Torsion Angles

The three-dimensional structure of any molecule, including a macromolecule, is uniquely specified by:

1. A diagram of the pattern of chemical bonds (the primary structure)

2. A list of the bond distances and bond angles

3. A list of the bond torsion angles (dihedral angles)

The bond distances and bond angles between specific atoms vary only slightly among different molecules. Thus, given a primary structure, the list of torsion angles determines the three-dimensional shape of a molecule.

Consider two adjacent amino acid residues, Gly-Asn, in a polypeptide chain. The three-dimensional structure of the dipeptide segment is determined by the six torsion angles, $\phi_1, \phi_2, \psi_1, \psi_2, \chi_1, \chi_2$ (Figure 7.1).

Figure 7.1

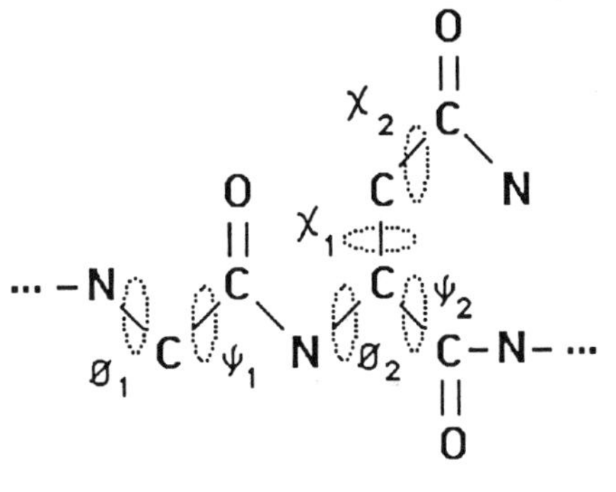

The three-dimensional structure of the dipeptide segment, Gly-Asn, can be defined only if six internal rotation angles (torsion angles) are known. Peptide bonds are assumed to be resonance stabilized into a planar configuration with the alpha carbons generally positioned *trans* with respect to the peptide bond. Hydrogen atoms are not displayed.

Chapter 7   Experiment 7.1

Because every amino acid except glycine has at least three single bonds about which rotation can occur, the three-dimensional structure of even a moderately sized protein can be represented only after several thousand torsion angles are known.

Fortunately, the seemingly endless number of combinations of torsion angles (and three-dimensional structures) is severely limited by the fact that no two atoms can occupy the same place at the same time. Analysis of the restrictions in allowed torsion angles due to steric hindrance was pioneered by Ramachandran.

Consider the simplest possible model of four atoms defining a torsion angle. Atoms A, B, C, and D are arranged such that the bond distances A-B, B-C, and C-D are equal, and the bond angles A-B-C and B-C-D are equal (Figure 7.2).

Figure 7.2

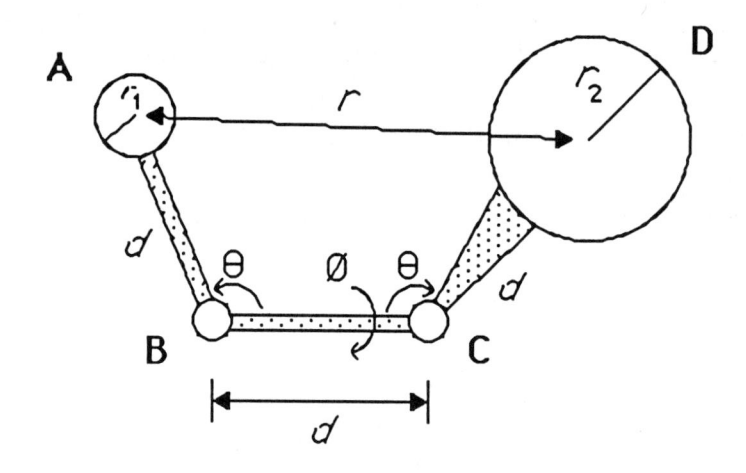

d = bond distance
θ = bond angle
r_1 = radius of atom A, a hard sphere
r_2 = radius of atom D, a hard sphere
ϕ = torsion angle
r = distance between the centers of atom A and atom D

Restrictions on Torsion Angles

The torsion angle ø is given by the angle between the plane specified by the atoms A, B, C and the plane specified by the atoms B, C, D. ø equals zero when all four atoms lie in the same plane, and atoms A and D are *cis* with respect to the B-C bond. ø increases as the bond C-D rotates clockwise with respect to the A-B bond as seen by an observer looking down the B-C axis from the B end of the molecule. (The atoms defining the dihedral angle must be specified explicitly if atoms B and C are bound to additional groups besides A and D.)

In this simplest case, r, the distance between the centers of atoms A and D, can be calculated from the equation:

$$r = d \sqrt{3 - 4\cos\theta + 2\cos^2\theta - 2\sin^2\theta \cos\emptyset} \qquad (7.1)$$

Because atoms A and D cannot occupy the same space at the same time, ø is restricted to values such that $r \geq r_1 + r_2$ (Figure 7.3).

Figure 7.3

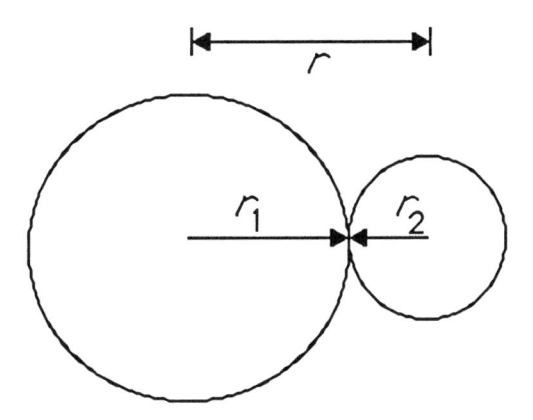

If r is the distance between the centers of two hard spheres with radii r_1 and r_2 (which cannot interpenetrate), r must be greater than or equal to $r_1 + r_2$.

Chapter 7   Experiment 7.1

Directions

1. Build and verify the models described in Figures 7.4 and 7.5. The models calculate the distance between atoms A and D in Figure 7.2.

2. Cells B5 and B6 in Figures 7.4 and 7.5 act as "lookup tables" for the values of $\sin(\theta)$ and $\cos(\theta)$. This arrangement increases the speed of calculation for these models.

3. The models assume that your electronic spreadsheet program requires that the arguments of trigonometric functions be expressed in radians. If this is not the case, delete the characters *PI/180 from these cells.

4. Figures 7.4 and 7.5 describe models of the same event. Figure 7.4 permits precise control of the value of the angle ø whereas Figure 7.5 assumes a specific range of values and automatically generates a table of values of ø and r. If ø is allowed (i.e., $r > r_1 + r_2$), ø is marked with a 1. If ø is not allowed (i.e., $r \leq r_1 + r_2$), ø is marked with a 0. Use either model to answer the questions that follow.

Figure 7.4

Formulas

	A	B
1	d =	1.5
2	theta =	109
3	phi =	20
4		
5	sin(theta) =	=SIN(B2*PI/180)
6	cos(theta) =	=COS(B2*PI/180)
7		
8	r =	***

Values

	A	B
1	d =	1.5
2	theta =	109
3	phi =	20
4		
5	sin(theta) =	0.9455
6	cos(theta) =	-0.3256
7		
8	r =	2.5252

Cell B8 is =B1*SQRT(3-(4*B6)+(2*B6*B6)-(2*B5*B5*COS(B3*PI/180)))

Restrictions on Torsion Angles

Figure 7.5

Formulas

	A	B	C	D	E
1	d =	1.5			
2	theta =	109			
3	r1 + r2 =	3			
4			phi	r	
5	sin(theta) =	=SIN(B2*PI/180)	180		=IF(D5>=B3,1,0)
6	cos(theta) =	=COS(B2*PI/180)	135		=IF(D6>=B3,1,0)
7			90	Formula Set I	=IF(D7>=B3,1,0)
8			45		=IF(D8>=B3,1,0)
9			0		=IF(D9>=B3,1,0)
10			-45		=IF(D10>=B3,1,0)
11			-90		=IF(D11>=B3,1,0)
12			-135		=IF(D12>=B3,1,0)
13			-180		=IF(D13>=B3,1,0)

Values

	A	B	C	D	E
1	d =	1.5			
2	theta =	109			
3	r1 + r2 =	3			
4			phi	r	
5	sin(theta) =	0.9455	180	3.77	1
6	cos(theta) =	-0.3256	135	3.61	1
7			90	3.19	1
8			45	2.70	0
9			0	2.48	0
10			-45	2.70	0
11			-90	3.19	1
12			-135	3.61	1
13			-180	3.77	1

Formula Set I
Prototype Cell is D5
=**B1***SQRT(3-(4***B6**)+(2***B6***B6**)-(2***B5***B5***COS(C5*PI/180)))

Chapter 7   Experiment 7.1

Questions

1. Complete the table below:

$r_1 + r_2 = 3$ Å $\theta = 109°$

	d = 1.5 Å			d = 1.75 Å		d = 2.0 Å	
ø	r	ø Allowed?	r	ø Allowed?	r	ø Allowed?
180	3.766	Yes	___	___	___	___
135	3.606	Yes	___	___	___	___
90	3.187	Yes	___	___	___	___
45	2.704	No	___	___	___	___
0	2.477	No	___	___	___	___
-45	2.704	No	___	___	___	___
-90	___	___	___	___	___	___
-135	___	___	___	___	___	___
-180	___	___	___	___	___	___

2. As the bond distance d increases does the allowable range of ø increase or decrease?

3. As the bond angle θ increases does the allowable range of ø increase or decrease? (You will need to design a separate table to answer this question.)

4. When the bond distance d equals 1.5 Å, what is the smallest value of the bond angle θ that allows atom A (Figure 7.2) to freely rotate a full 360° about the B-C bond? Use $r_1 + r_2 = 3$ Å.

5. The fraction of all ø values in the allowed region provides an estimate of the degree of movement available to a group. If d = 1.5 Å and θ = 109°, how much does this fraction change as r_1 increases from 1.7 Å to 2.0 Å, keeping r_2 fixed at 1.7 Å? How might this information be used to lock a molecule into a narrow range of conformations?

Restrictions on Torsion Angles

Problems

1. The models described in Figures 7.4 and 7.5 were simplified by forcing equality of all bond distances and bond angles. Write a model for the behavior of four atoms A, B, C, and D where these simplifying assumptions no longer exist (Figure 7.6).

Figure 7.6

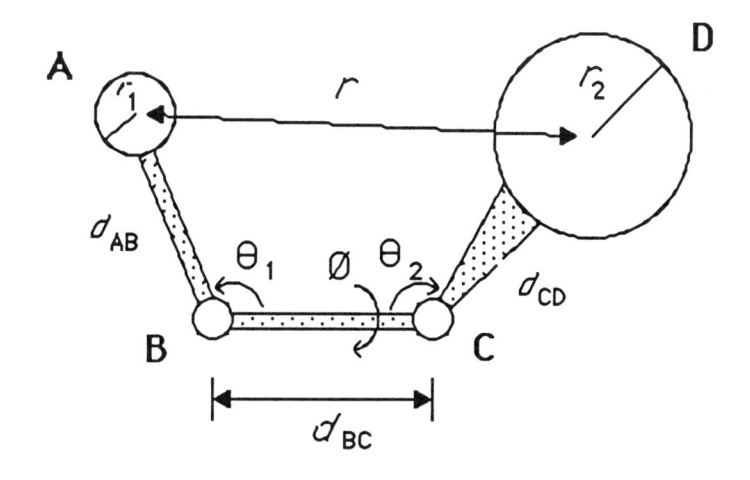

d_{AB}	=	bond distance A-B
d_{BC}	=	bond distance B-C
d_{CD}	=	bond distance C-D
θ_1	=	bond angle, ABC
θ_2	=	bond angle, BCD
r_1	=	radius of atom A, a hard sphere
r_2	=	radius of atom D, a hard sphere
$\emptyset$	=	torsion angle
r	=	distance between the centers of atom A and atom D

The distance between the centers of atom A and atom D, r, is calculated using the Equation 7.2:

$$r = \sqrt{d_{AB}^2 + d_{BC}^2 + d_{CD}^2 - 2d_{AB}d_{BC}\cos\theta_1 - 2d_{BC}d_{CD}\cos\theta_2 + 2d_{AB}d_{CD}(\cos\theta_1\cos\theta_2 - \sin\theta_1\sin\theta_2\cos\emptyset)}$$

(7.2)

Tables 1 through 3 at the end of this experiment give the van der Waals radii, bond distances, and bond angles for selected atoms and bonds. Use your model and the information in these tables to find the range of allowable values of ø for the following structures:

$$\begin{array}{ccc} \text{C} \quad\quad \text{C} & \text{N} \quad\quad \text{N} & \text{C} \quad\quad \text{C}_\beta \\ \diagdown \diagup & \diagdown \diagup & \diagdown \diagup \\ \text{N}-\text{C}_\alpha & \text{C}_\alpha-\text{C} & \text{N}-\text{C}_\alpha \\ \text{(a)} & \text{(b)} & \text{(c)} \end{array}$$

2. Real atoms are not hard spheres. They deform slightly in the presence of other atoms. This softness can be modeled using a Lennard-Jones potential. The interaction energy E of two approaching atoms may be calculated using the equation:

$$E = \frac{B}{r^{12}} - \frac{A}{r^6} \quad\quad (7.3)$$

Where E = interaction energy
r = distance between two atoms in Å
B = first coefficient
A = second coefficient

For two interacting carbon atoms:
$B = 2.75 \times 10^6$ kcal Å^{12}/mol
$A = 1425$ kcal Å^6/mol

Build a simple model (all bond distances and angles equal) in which the interaction energy between carbon atoms may not exceed RT (where R is the gas constant 1.98 cal/°K, and T is the absolute temperature). At room temperature RT is approximately 0.59 kcal/mol.

RT provides a convenient approximation of the upper limit for the interaction energy because, according to the Boltzmann distribution (Experiment 4.1), interactions with energies significantly exceeding RT are unlikely to occur.

If all bond distances equal 1.5 Å and all bond angles are 109°, what are the allowed torsion angles ø of the model? For $d = 1.75$ Å? $d = 2.0$ Å?

Restrictions on Torsion Angles

3. To develop a more realistic model for carbon atoms, you must extend your treatment to cases in which atoms B and C (Figure 7.2) have more than one substituent. Saturated carbon atoms have four substituents, organized in the familiar tetrahedral arrangement:

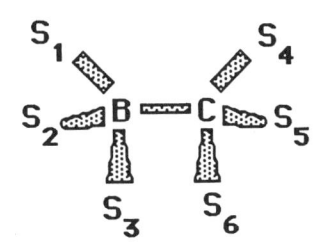

To determine the allowed region of ø for this case, the distances between all possible substituent pairs (i.e., S_1-S_4, S_1-S_5, S_1-S_6, S_2-S_4, S_2-S_5, S_2-S_6, S_3-S_4, S_3-S_5, S_3-S_6) must be considered. For carbon atoms with ideal tetrahedral symmetry, the extension is relatively straightforward. If atoms S_1, B, C, S_4 describe the torsion angle ø, then atoms S_1, B, C, S_6 have the torsion angle ø + 120°, and atoms S_1, B, C, S_5 have the torsion angle ø − 120°. The other torsion angles are obtained in a similar fashion.

With the values of all torsion angles known, the distances between each of the substituents can be calculated. Extend your model to include these additional angles and calculate the allowed region for ø for different values of hard sphere radii and bond distances. What happens to the range of allowed values of ø with these additional assumptions in place? What are the implications of this behavior for protein folding?

Chapter 7   Experiment 7.1

Table 7.1

van der Waals Radii

Type of Atom	van der Waals Radius (Å)
O	1.4
N	1.7
C	1.9
S	1.9

Table 7.2

Bond Distances

Type of Bond	Bond Distance (Å)
$C_\alpha - C$	1.53
$C - N$	1.32
$N - C_\alpha$	1.47
$C = O$	1.24
$C_\alpha - C_\beta$	1.54

Table 7.3

Bond Angles

Type of Bond	Bond Angle (°)
$C_\alpha - C - N$	113°
$C - N - C_\alpha$	123°
$N - C_\alpha - C$	110°
$O = C - N$	123°
$N - C_\alpha - C_\beta$	109°

Experiment 7.2*

Three-Dimensional Structures of Macromolecules:

Restrictions on Two Consecutive Torsion Angles

The torsion angles of three-dimensional molecular structures are restricted to certain values by the fact that no two atoms can exist in the same place at the same time. In Experiment 7.1 such restrictions were a function of:

1. The primary structure of the molecule
2. The bond distances and bond angles of the molecule
3. The effective radii of the atoms in the molecule

In fact, torsion angles are also restricted by the values of other torsion angles. Consider the simplest possible example of a molecule with two torsion angles, $\emptyset_1$ and $\emptyset_2$ (Figure 7.7).

Figure 7.7

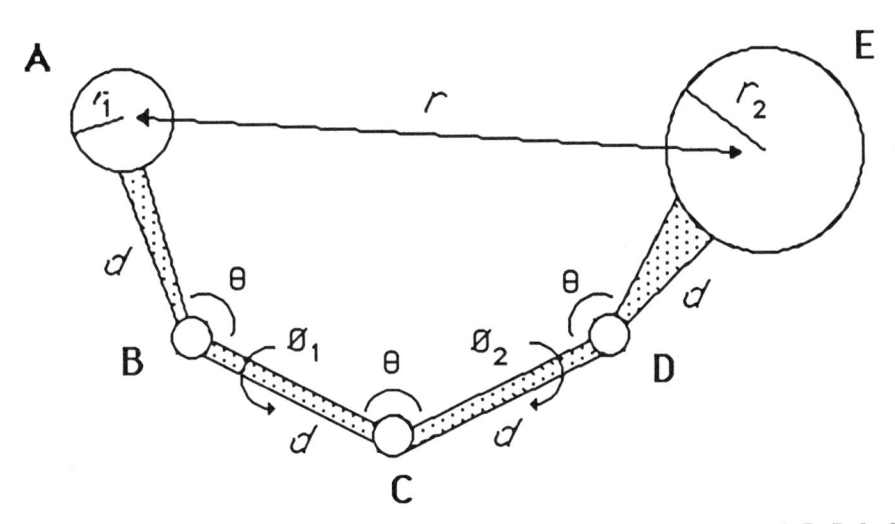

Atoms A, B, C, D, and E are linked together such that all bond distances (A-B, B-C, C-D, and D-E) and all bond angles (ABC, BCD, and CDE) are equal. The radii of atoms B and D are presumed to be negligible thus eliminating consideration of A-D or B-E interactions.

277

Chapter 7   Experiment 7.2*

Figure 7.7 (Contd.)

d = common bond distance
θ = common bond angle
ϕ_1 = torsion angle about bond B-C
ϕ_2 = torsion angle about bond C-D
r_1 = radius of atom A, a hard sphere
r_2 = radius of atom E, a hard sphere
r = distance between the centers of atom A and atom E

In this case, r, the distance between the centers of atoms A and E, can be calculated from the equation:

$$r = d \sqrt{\left(4 \sin^3 \tfrac{\theta}{2} - 2 \sin\theta \cos\tfrac{\theta}{2} \cos\alpha \cos\beta\right)^2 + \left(2 \sin\theta \sin\tfrac{\theta}{2} \sin\alpha \sin\beta\right)^2 + \left(2 \sin\theta \sin\alpha \cos\beta\right)^2} \qquad (7.4)$$

where $\quad \alpha = \dfrac{\phi_1 + \phi_2}{2} \quad\quad \beta = \dfrac{\phi_1 - \phi_2}{2}$

Thus r depends on both ϕ_1 and ϕ_2. As a result, allowed values of ϕ_1 are affected by the value of ϕ_2, and vice-versa.

Directions

Build and verify the model described in Figure 7.8. The model calculates the allowed values of ϕ_1 and ϕ_2 based on the separation between atoms A and E.

Restrictions on Two Consecutive Torsion Angles

Figure 7.8

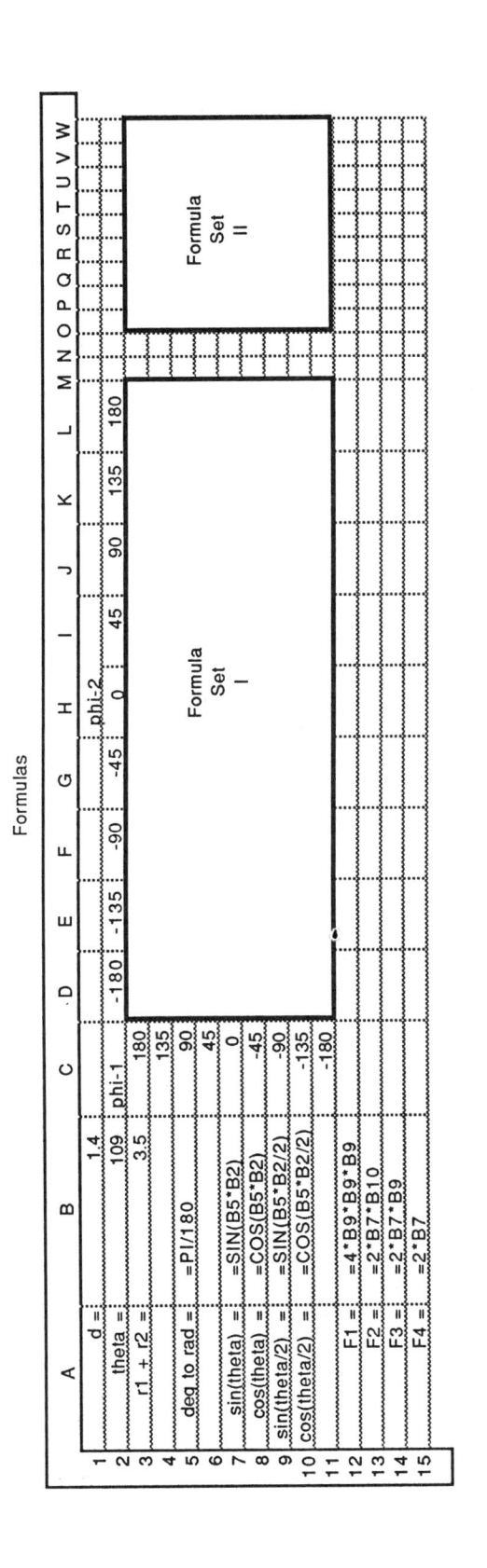

	A	B	C	D	E	F	G	H	I	J	K	L	M N O P Q R S T U V W
1													
2	d =	1.4											
3	theta =	109	phi-1	-180	-135	-90	-45	phi-2 0	45	90	135	180	
4	r1 + r2 =	3.5	180										Formula Set II
5	deg_to_rad =	=PI/180	135										
6			90					Formula					
7	sin(theta) =	=SIN(B5*B2)	45					Set					
8	cos(theta) =	=COS(B5*B2)	0					I					
9	sin(theta/2) =	=SIN(B5*B2/2)	-45										
10	cos(theta/2) =	=COS(B5*B2/2)	-90										
11			-135										
12	F1 =	=4*B9*B9*B9	-180										
13	F2 =	=2*B7*B10											
14	F3 =	=2*B7*B9											
15	F4 =	=2*B7											

Formulas

(Continued.)

279

Chapter 7   **Experiment 7.2***

Figure 7.8 (Contd.)

	A	B	C	D	E	F	G	H	I	J	K	L	M	N	O	P	Q	R	S	T	U	V	W
1		1.4																					
2	d =	109	phi-1	-180	-135	-90	-45	phi-2 0	45	90	135	180											
3	theta =	3.5		180	135	90	45	0	-45	-90	-135	-180		1	1	1	1	1	1	1	1	1	1
4	r1 + r2 =			4.56	4.45	4.16	3.85	3.71	3.85	4.16	4.45	4.56		1	1	1	0	1	1	1	1	1	1
5	deg to rad =	0.0175		4.45	4.11	3.67	3.38	3.48	3.38	3.67	4.11	4.45		1	1	0	0	0	0	0	1	1	1
6				4.16	3.67	3.02	2.62	2.83	2.62	3.02	3.67	4.16		1	0	0	0	0	0	0	0	1	1
7	sin(theta) =	0.9455		3.85	3.38	2.62	1.93	1.97	1.48	1.97	2.69	3.44	3.85		1	0	0	0	0	0	0	1	1
8	cos(theta) =	-0.3256		3.71	3.48	2.83	1.97	1.48	1.97	2.83	3.48	3.71		1	0	0	0	0	0	0	0	1
9	sin(theta/2) =	0.8141		3.85	3.87	3.44	2.69	1.97	2.69	3.44	3.87	3.85		1	1	0	0	0	0	0	1	1
10	cos(theta/2) =	0.5807		4.16	4.29	4.02	3.44	2.83	3.44	4.02	4.29	4.16		1	1	1	0	0	0	1	1	1
11				4.45	4.52	4.29	3.87	3.48	3.87	4.29	4.52	4.45		1	1	1	1	0	1	1	1	1
12	F1 =	2.1583		4.56	4.45	4.16	3.85	3.71	3.85	4.16	4.45	4.56		1	1	1	1	1	1	1	1	1
13	F2 =	1.0981																					
14	F3 =	1.5395																					
15	F4 =	1.8910																					

Values

Formula Set I
Prototype Cell is D3
=**B1***SQRT((**B12**-**B13***COS(**B5***(**C**3+**D2**)/2)*COS(**B5***(**C**3-**D2**)/2))^2+(**B14***SIN(**B5***(**C**3+**D2**)/2)*SIN(**B5***(**C**3-**D2**)/2))^2+(**B15***SIN(**B5***(**C**3+**D2**)/2)*COS(**B5***(**C**3-**D2**)/2))^2)

Formula Set II
Prototype Cell is O3
=IF(D3>**B3**,1,0)

Questions

1. The model described in Figure 7.8 explores allowable values of ϕ_1 and ϕ_2 in increments of 45°. There are 81 possible combinations of ϕ_1 and ϕ_2 displayed. What fraction of these 81 combinations are allowed when the bond distances in the model equal 1.5 Å, the bond angles equal 110°, and the sum of the radii of atoms A and E are 2.75 Å?

   What happens to the fraction of allowed values of ϕ_1 and ϕ_2 as the bond distances increase? As the sum of the radii of atoms A and E decrease? As the bond angles decrease?

2. This is one of those models that yields the greatest benefits to independent explorers. Do it.

Problems

1. The model described in Figure 7.8 ignores the contacts that might occur between atom pairs A-D and B-E. The separation distance between these atoms depends only on ϕ_1 (for A-D) and ϕ_2 (for B-E). These distances can be calculated using the model developed in Figure 7.5.

   Rewrite the model described in Figure 7.8 to include the effect of contacts between atom pairs A-D and B-E. How much conformational space is lost when these constraints are added to the model (assume that atoms A, B, D, and E all have a radius of 1.4 Å)?

2. For certain values of ϕ_2 the allowed region of ϕ_1 is not influenced by the presence of atom E (given the additional constraints imposed by the previous question). What values of ϕ_2 have no influence on the allowable range of ϕ_1? What values of ϕ_2 have the greatest influence on the allowable range of ϕ_1?

3. With the constraints added by Question 1, your model now resembles the behavior of a real polypeptide chain. For example, atoms A, B, C, D, and E may be taken to correspond to the atoms C,

Chapter 7 Experiment 7.2*

N, C_α, C, and N of a polyglycine chain. In this case, the angles ϕ_1 and ϕ_2 would correspond to the peptide chain dihedral angles ϕ and ψ:

```
                                    O
                                    ‖
    A       E           C       N
    |       |    ≡      |       |
    B       D           N       C
     \     /              ↶  ↷    \
      \   /                ϕ   ψ    \
       C                  Cα         O
```

Restrictions on ϕ and ψ generate a Ramachandran diagram for polyglycine. By researching the matter in other texts, point out the regions of your model's simplified Ramachandran diagram corresponding to the familiar secondary structures of alpha helices and beta sheets.

Experiment 7.3

Electrostatic Influences on Macromolecular Structure: General Considerations

Electrostatic interactions are the predominant long-range force in biochemical systems. They play an essential role in stabilizing macromolecular structure and in promoting specific interactions between ligands and macromolecules.

The fundamental expression of electrostatics is given by Coulomb's law (Equation 7.5). Coulomb's law calculates the potential energy of interaction between two charged bodies embedded in a medium of known dielectric constant (Figure 7.9):

Figure 7.9

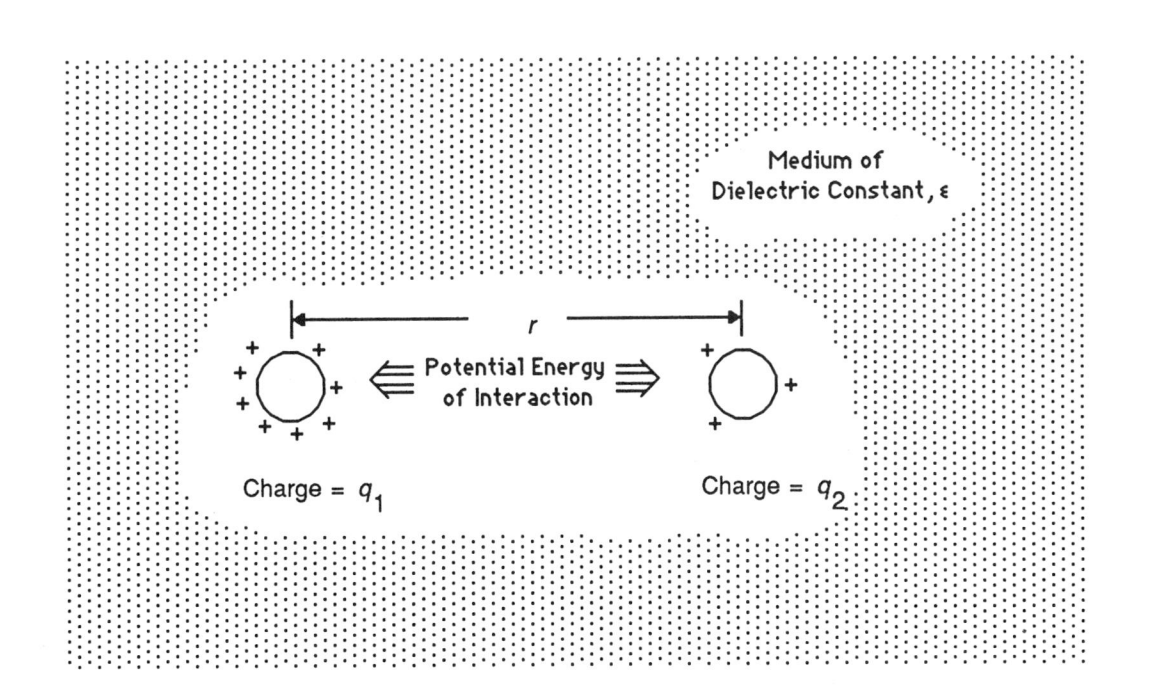

Chapter 7   Experiment 7.3

$$V = \frac{332\, q_1 q_2}{\varepsilon\, r} \tag{7.5}$$

where V = the potential energy of interaction between two charged bodies (in kcal/mol)
q_1 = charge on the first body (in multiples of the protonic charge)
q_2 = charge on the second body (in multiples of the protonic charge)
ε = dielectric constant of the surrounding medium
r = separation distance between the two bodies (in Å)

In addition to the dielectric effects of the solvent, electrostatic interactions are dampened by the presence of salts in solution. The bodies with charges q_1 and q_2 will be surrounded by a cloud of oppositely charged ions that effectively screen the electrostatic potential (Figure 7.10).

Figure 7.10

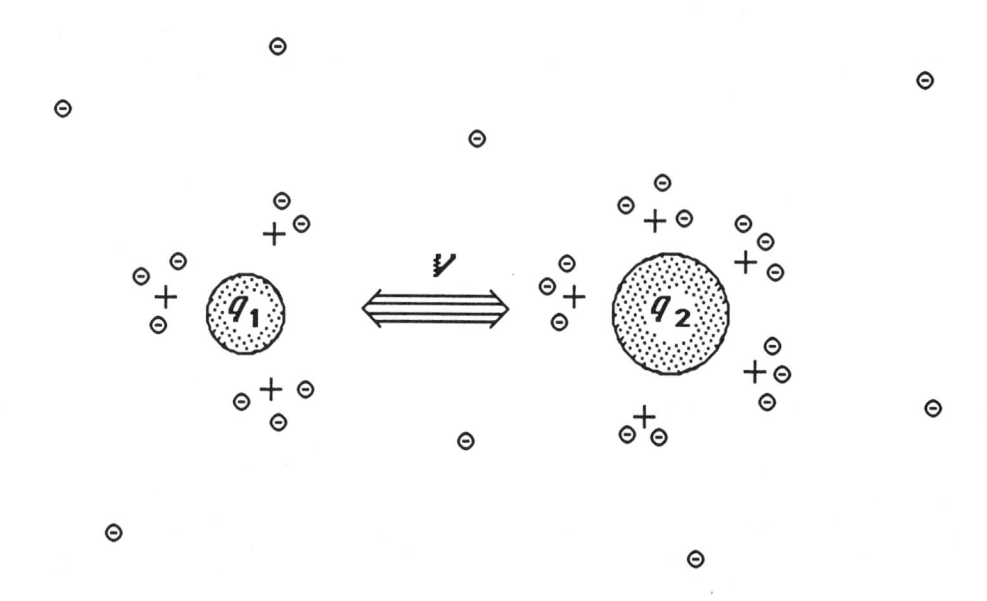

Electrostatic Influences on Macromolecular Structure

In this instance the potential energy of interaction can be calculated from a limiting expression first derived by Debye and Huckel:

$$V = \frac{332 q_1 q_2}{\varepsilon r} e^{-(0.332 \sqrt{I}\, r)} \qquad (7.6)$$

Where I = ionic strength (M)

This expression is the Coulomb formula, modified by an exponential screening term.

Directions

Build and verify the models described in Figures 7.11 and 7.12.

Figure 7.11

Formulas

	A	B
1	q1 =	1
2	q2 =	1
3	r =	9
4	dielectric const. =	80
5		
6	V =	=332*B1*B2/(B4*B3)

Values

	A	B
1	q1 =	1
2	q2 =	1
3	r =	9
4	dielectric const. =	80
5		
6	V =	0.4611

Chapter 7   Experiment 7.3

Figure 7.12

Formulas

	A	B
1	q1 =	1
2	q2 =	1
3	r =	9
4	dielectric const. =	80
5	I =	0.01
6		
7	V =	=(332*B1*B2/(B4*B3))*EXP(-0.332*SQRT(B5)*B3)

Values

	A	B
1	q1 =	1
2	q2 =	1
3	r =	9
4	dielectric const. =	80
5	I =	0.01
6		
7	V =	0.3420

Questions

1. Using the model described in Figure 7.11, complete and graph the table below by finding the potential energy of interaction for the combinations of dielectric constants and distances shown. Assume that the electrostatic interactions are those of two sodium ions ($q_1 = q_2 = 1$).

Potential Energy as a Function of ε and r

		Dielectric Constant		
		10	40	80
D	2	___	___	___
i	4	___	___	___
s	6	___	___	___
t	8	___	___	___
a	10	___	___	___
n	12	___	___	___
c	14	___	___	___
e	16	___	___	___
	18	___	___	___

Electrostatic Influences on Macromolecular Structure

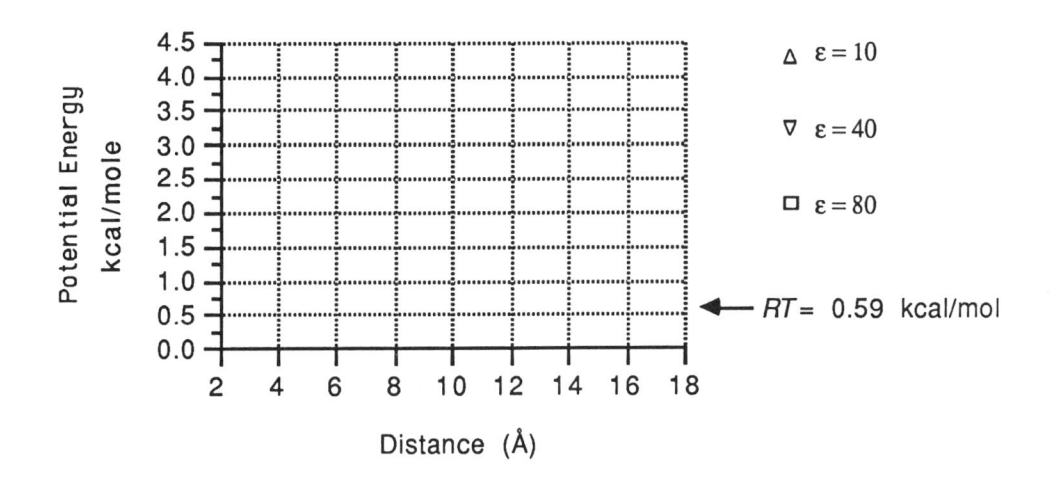

[Graph: Potential Energy (kcal/mole) vs Distance (Å), with curves for Δ ε = 10, ▽ ε = 40, □ ε = 80, and a marker for RT = 0.59 kcal/mol]

What values of r and ε yield potential energies of interaction that exceed the average thermal energy RT at 20°C?

2. Compare the distance dependence of the potential energy of interaction due to electrostatic forces with the distance dependence of the Lennard-Jones potential (Experiment 7.1) by completing and graphing the table below.* Of the two potential energies, which is the long-range interaction?

Distance (Å)	Electrostatic Potential Energy (kcal/mole)	Lennard-Jones Potential (kcal/mol)
2.0	_____	_____
3.0	_____	_____
4.0	_____	_____
5.0	_____	_____
6.0	_____	_____

* Consider the interaction of two sodium ions in water ($\varepsilon = 80$) and neglect ionic strength effects. For purposes of this calculation, model the Lennard-Jones potential for sodium with the A and B coefficients for carbon given in Experiment 7.1.

Chapter 7   Experiment 7.3

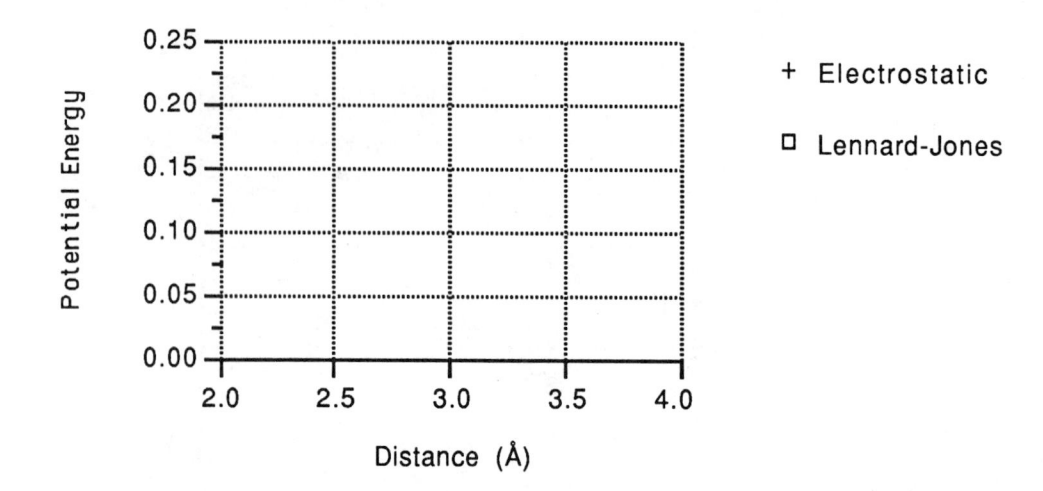

3. Using the model described in Figure 7.12, calculate the electrostatic potential energy of interaction between two sodium ions separated by 10 Å in solutions of $0M$, $0.001M$, $0.01M$, and $0.1M$ ionic strength. Assume a polar environment in which $\varepsilon = 80$. What if the sodium ions are separated by 30 Å? 100 Å?

a. Interactions between many biomolecules significantly depend on ionic strength. How might increasing ionic strength influence the binding between a positively charged repressor molecule and negatively charged DNA?

b. How might increasing ionic strength influence the binding between a negatively charged flavodoxin molecule and a positively charged ion exchange resin? These effects are of major importance for the purification of macromolecules.

	Electrostatic Potential Energy		
Ionic Strength (M)	10 Å	30 Å	100 Å
0	___	___	___
0.001	___	___	___
0.01	___	___	___
0.1	___	___	___

Electrostatic Influences on Macromolecular Structure

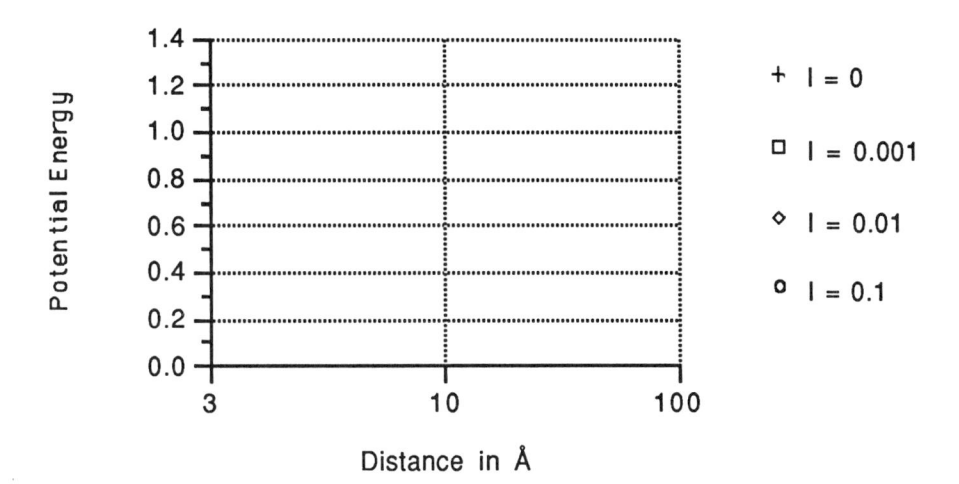

Problems

1. Devise a single model capable of investigating the effects of r, I, and ε on electrostatic energies. Consider the structure below as a starting point.

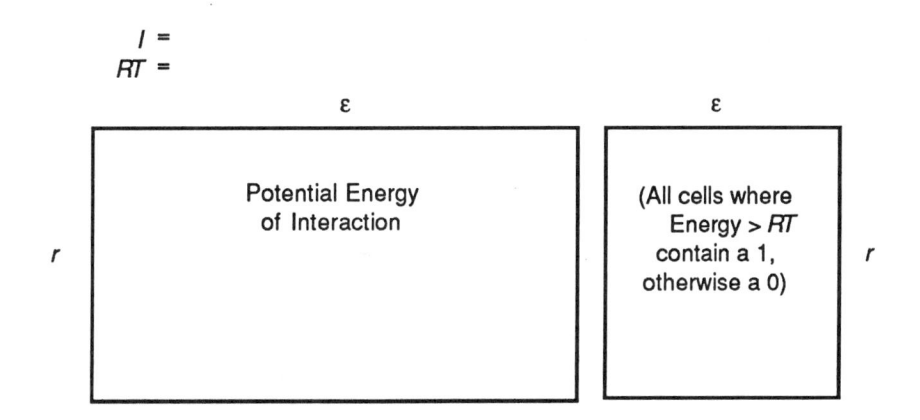

2. Experiment 3.5 explored the influence of pH on the distribution of an ionizable molecule between an aqueous and an organic medium. Generally, the charged form of an ionizable group is more stable in an aqueous environment (and the less charged group more stable in a nonpolar environment) due to electrostatic effects.

Chapter 7 Experiment 7.3

The contribution of electrostatic forces to the free energy of a charged molecule may be roughly estimated by the Born expression:

$$G = \frac{332 q^2}{2 \varepsilon r_p} \tag{7.7}$$

where
- q = charge on the molecule
- r_p = radius of the ion in Å
- G = free energy in kcal/mol
- ε = dielectric constant

As a result, the change in free energy associated with transferring a charged group from a polar to a nonpolar environment (with dielectric constants ε_P and ε_N, respectively), may be approximated by:

$$G_e = \frac{332 q^2}{2 r_p}\left(\frac{1}{\varepsilon_N} - \frac{1}{\varepsilon_P}\right) \tag{7.8}$$

The free-energy change ΔG_e is designated with a subscript e to emphasize that this calculation refers only to the electrostatic contribution to group transfer — the only contribution considered in this experiment. ΔG_e provides a major, although not exclusive, contribution to the thermodynamics of transfer.

Because ε_N must be less than ε_P, $1/\varepsilon_N - 1/\varepsilon_P$ is greater than zero, and the electrostatic contribution to this process is always thermodynamically unfavorable. If the group is electrically neutral, $q = 0$ and $\Delta G_e = 0$.

The change in free energy associated with protonating an ionizable group is computed from Equation 7.9:

$$\Delta G_H = 2.303 RT(\text{pH} - \text{pK}) \tag{7.9}$$

Equations 7.8 and 7.9 allow you to examine the distribution of the charged and uncharged forms of a group between a polar and a nonpolar (organic) environment (Figure 7.13).

Figure 7.13

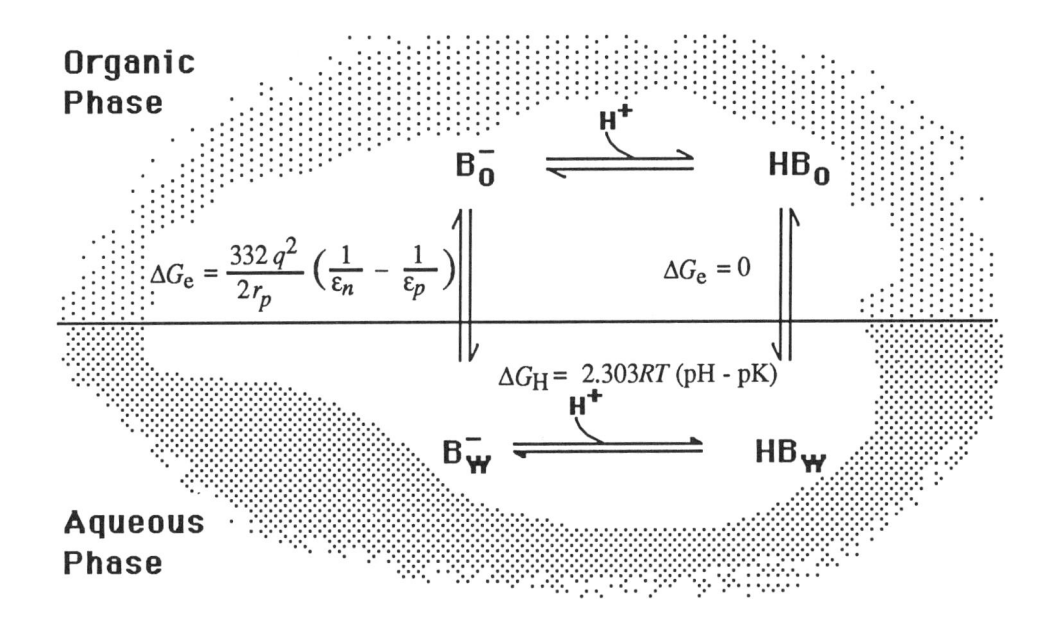

$$\Delta G_e = \frac{332 q^2}{2r_p}\left(\frac{1}{\varepsilon_n} - \frac{1}{\varepsilon_p}\right)$$

$\Delta G_e = 0$

$\Delta G_H = 2.303 RT (pH - pK)$

The distribution of the charged and uncharged forms of a group between a polar and organic phase can be calculated by using the relationships described in Equations 7.8 and 7.9. ΔG_e is the change in electrostatic free energy associated with transferring a charged group from a polar to a nonpolar environment. ΔG_H is the change in free energy associated with protonating an ionizable group in an aqueous phase. The free energy associated with protonating an ionizable group in the organic phase is implicitly specified by the other three events identified in the figure.

This situation outlined in Figure 7.13 is analogous to the partitioning problem explored in Experiment 3.5. In Experiment 3.5 the distribution of conjugate acid and conjugate base between the two phases was described in terms of the pK, pH, and solubilities of the different components.

The basic partition expression for this system was given in Experiment 3.5 as Equation 3.23 and is repeated here as Equation 7.10.

$$100 = [HB]_w \left[1 + \frac{[HB]_o}{[HB]_w} + \frac{[B]_w}{[HB]_w}\left(1 + \frac{[B]_o}{[B]_w}\right)\right] \quad (7.10)$$

Equation 7.11 repeats Equation 3.24 in Experiment 3.5 and results from the fact that $[B]_w/[HB]_w$ may be computed from Equation 7.9

Chapter 7   Experiment 7.3

(or the Henderson-Hasselbalch equation) and that $[HB]_o/[HB]_w$ and $[B]_o/[B]_w$ are determined from the solubilities of each component in the organic and aqueous phases.

$$100 = [HB]_w \left[1 + \frac{S[HB]_o}{S[HB]_w} + 10^{(pH-pK)} \left(1 + \frac{S[B]_o}{S[B]_w} \right) \right] \quad (7.11)$$

(Note: Equations 3.23, 3.24, 7.10, and 7.11 assume that the total amount of conjugate acid and conjugate base is 100 mmoles.)

The relative solubilities of the different components can be estimated from first principles using Equation 7.8:

$$\frac{S[HB]_o}{S[HB]_w} = e^{-0/RT} = 1 \quad (7.12)$$

$$\frac{S[B_o]}{S[B_w]} = e^{\left[-\frac{332 q^2}{2 r_p RT} \left(\frac{1}{\varepsilon_N} - \frac{1}{\varepsilon_P} \right) \right]} \quad (7.13)$$

Remember, however, the assumption that solubility ratios are determined exclusively by electrostatic forces is naive.

Nevertheless, these effects are important in the transfer process. You should design and build a model based on Figure 7.13, which calculates the distribution of the conjugate acid and conjugate base forms of a carboxylic acid between organic and aqueous phases as a function of pH.

Assume that $pK = 4.8$, $\varepsilon_P = 80$, $r_p = 4$ Å and $T = 198°K$. Calculate these distributions as the pH varies from 2 to 12, and ε_N varies from 2 to 80.

3. In Problem 2 the free energy of ionization of BH in the nonpolar phase is given by $\Delta G_H - \Delta G_e$. The apparent pK of BH, pK', in the nonpolar environment can be obtained by finding the pH of the aqueous solution for which $\Delta G_H - \Delta G_e = 0$ and thus:

$$\text{pK}' = \frac{332\, q^2}{2(2.303) r_p RT} \left(\frac{1}{\varepsilon_N} - \frac{1}{\varepsilon_P} \right) + \text{pK}$$

Build a model that computes pK' from pK, ε_N, and r_p. Hold the dielectric constant of the polar environment constant at $\varepsilon_P = 80$ and the temperature at 27°C.

Experiment 7.4

Electrostatic Influences on Macromolecular Structure:
pH Dependence of Protein Stability

Electrostatic interactions are a principal component of the pH dependence of protein stability. As noted in Equation 7.5 (Problem 1, Experiment 7.3), the electrostatic work required to place a total charge of q on a spherical protein may be approximated by:

$$G = \frac{332 q^2}{2 \varepsilon r_p} \tag{7.14}$$

where G = work required to charge the protein (kcal/mol)
q = charge to be added
ε = dielectric constant
r_p = radius of the protein accepting the charge (Å)

Because G depends on q^2 rather than q, G is always positive, and work must be performed to charge up the protein. The amount of work required is inversely proportional to the radius of the protein accepting the charge. When the protein is unfolded and the charges are far apart (i.e., the radius is large), less work is required than when the protein is compact and the charges are close together.

If a completely unfolded peptide chain is assumed to approximate the limit as $r_p \to \infty$ then the work G required to charge an unfolded protein is zero, and the overall electrostatic free-energy change ΔG_{ES}, for the process of folding a charged protein, is:

$$\Delta G_{ES} = \frac{332 q^2}{2 \varepsilon r_p} - 0 = \frac{332 q^2}{2 \varepsilon r_p} \tag{7.15}$$

As noted in Chapters 1 and 2, the net charge q of a protein depends on the number of acidic and basic side-chain-bearing amino acids and pH. If the charged groups do not interact with each other, the fraction of each residue in the deprotonated state can be given by:

$$\text{Fraction deprotonated} = \frac{10^{pH-pK}}{1 + 10^{pH-pK}} \tag{7.16}$$

Thus, if the amino acid composition of the protein is known and the pH is specified, the net charge on the protein and the corresponding electrostatic ΔG accompanying folding can be calculated using the relationships introduced in Chapter 2 (Experiment 2.4).

Finally, suppose that the total free energy of the folding process can be divided into two terms:

$$\Delta G_T = \Delta G_{NES} + \Delta G_{ES} \tag{7.17}$$

where ΔG_{NES} = nonelectrostatic contributions to the free energy of folding
ΔG_{ES} = electrostatic contribution to the free energy of folding

If ΔG_{NES} is independent of pH, then the total free energy of the folding reaction can be calculated as:

$$\Delta G_T = \Delta G_{NES} + \frac{332 q^2}{2 \varepsilon r_p} \tag{7.18}$$

where q is determined by finding the charges contributed by each amino acid side-chain (Equation 7.16), multiplying by the number of side-chains of that type, and summing all charges.

These considerations are incorporated into the following model.

Directions

1. Build and verify the model described in Figure 7.14.

2. The model is driven by entering the number of ionizable side-chain-bearing amino acids in column C and the charge of the acidic and basic forms of these side chains in Columns D and E, respectively. ΔG_T and the fraction of protein in the native

Chapter 7   Experiment 7.4

conformation are read in rows 13 and 14, respectively. Contributions of nonelectrostatic forces and the effective radius of the protein are entered in cells B2 and B1, respectively.

Figure 7.14

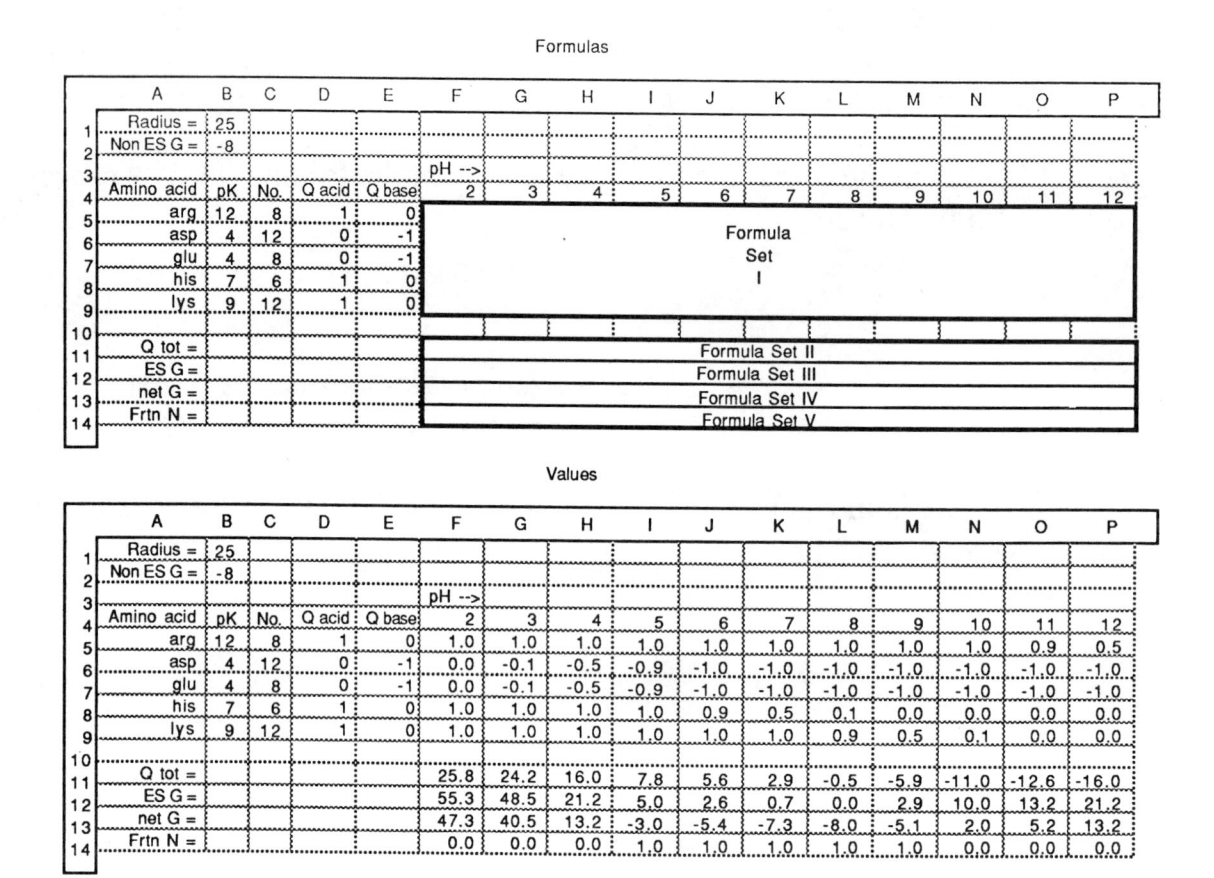

Formulas

	A	B	C	D	E	F	G	H	I	J	K	L	M	N	O	P
1	Radius =	25														
2	Non ES G =	-8														
3						pH -->										
4	Amino acid	pK	No.	Q acid	Q base	2	3	4	5	6	7	8	9	10	11	12
5	arg	12	8	1	0											
6	asp	4	12	0	-1					Formula						
7	glu	4	8	0	-1					Set						
8	his	7	6	1	0					I						
9	lys	9	12	1	0											
10																
11	Q tot =										Formula Set II					
12	ES G =										Formula Set III					
13	net G =										Formula Set IV					
14	Frtn N =										Formula Set V					

Values

	A	B	C	D	E	F	G	H	I	J	K	L	M	N	O	P
1	Radius =	25														
2	Non ES G =	-8														
3						pH -->										
4	Amino acid	pK	No.	Q acid	Q base	2	3	4	5	6	7	8	9	10	11	12
5	arg	12	8	1	0	1.0	1.0	1.0	1.0	1.0	1.0	1.0	1.0	1.0	0.9	0.5
6	asp	4	12	0	-1	0.0	-0.1	-0.5	-0.9	-1.0	-1.0	-1.0	-1.0	-1.0	-1.0	-1.0
7	glu	4	8	0	-1	0.0	-0.1	-0.5	-0.9	-1.0	-1.0	-1.0	-1.0	-1.0	-1.0	-1.0
8	his	7	6	1	0	1.0	1.0	1.0	1.0	0.9	0.5	0.1	0.0	0.0	0.0	0.0
9	lys	9	12	1	0	1.0	1.0	1.0	1.0	1.0	1.0	0.9	0.5	0.1	0.0	0.0
10																
11	Q tot =					25.8	24.2	16.0	7.8	5.6	2.9	-0.5	-5.9	-11.0	-12.6	-16.0
12	ES G =					55.3	48.5	21.2	5.0	2.6	0.7	0.0	2.9	10.0	13.2	21.2
13	net G =					47.3	40.5	13.2	-3.0	-5.4	-7.3	-8.0	-5.1	2.0	5.2	13.2
14	Frtn N =					0.0	0.0	0.0	1.0	1.0	1.0	1.0	1.0	0.0	0.0	0.0

(Continued.)

Figure 7.14 (Contd.)

Formula Set I
Prototype Cell is F5
=D5/(1+10^(F4-B5))+E5*(10^(F4-B5))/(1+10^(F4-B5))

Formula Set II
Prototype Cell is F11
=(C5*F5)+(C6*F6)+(C7*F7)+(C8*F8)+(C9*F9)

Formula Set III
Prototype Cell is F12
=332*F11*F11/(2*80*B1)

Formula Set IV
Prototype Cell is F13
=B2+F12

Formula Set V
Prototype Cell is F14
=EXP(-F13/0.6)/(1+EXP(-F13/0.6))

Chapter 7   Experiment 7.4

Questions

1. Complete and graph the tables below to explore how changes in the radius of a protein influences its denaturation curve. Assume that $\Delta G_{NES} = -8$ and that the amino acid composition is the same as in the initialized model (Figure 7.14).

Values of ΔG_T vs. pH for Different r_p Values

r_p	2	3	4	5	6	pH 7	8	9	10	11	12
15	__	__	__	__	__	__	__	__	__	__	__
25	__	__	__	__	__	__	__	__	__	__	__
50	__	__	__	__	__	__	__	__	__	__	__

Fraction of Native Protein vs. pH for Different r_p Values

r_p	2	3	4	5	6	pH 7	8	9	10	11	12
15	__	__	__	__	__	__	__	__	__	__	__
25	__	__	__	__	__	__	__	__	__	__	__
50	__	__	__	__	__	__	__	__	__	__	__

△ $r_p = 15$
▽ $r_p = 25$
□ $r_p = 50$

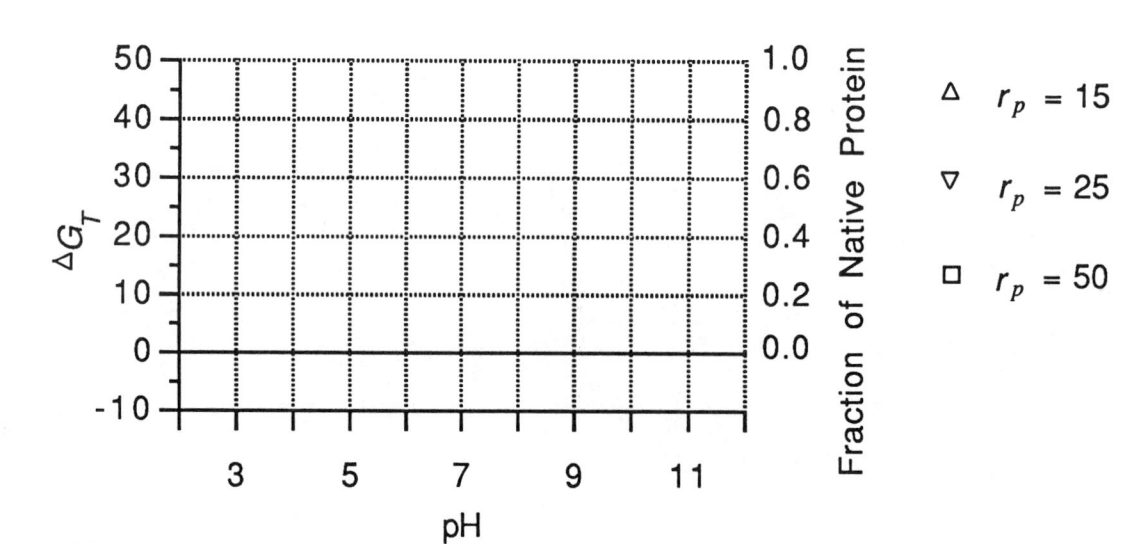

pH Dependence of Protein Stability

a. As the radius of the protein increases, what happens to ΔG_T at the pH value that promotes maximum stability? Why?

b. As the radius of the protein increases, what happens to ΔG_T at pH values that do not promote maximum stability? Why?

c. Obtain additional information from the model described in Figure 7.14 to determine range of pH values over which a protein with a radius of 25 Å remains at least 99% folded. What is the range of pH values if $r_p = 35$ Å?

2. Complete and graph the table below to explore the effect of increasing numbers of lysine residues on ΔG_T. Assume a radius of 25 Å. Give the other parameters the same values as in Figure 7.14.

Values of ΔG_T vs. pH for Different Numbers of Lysine Residues

# lys	2	3	4	5	6	7	8	9	10	11	12
4	__	__	__	__	__	__	__	__	__	__	__
8	__	__	__	__	__	__	__	__	__	__	__
12	__	__	__	__	__	__	__	__	__	__	__

△ # lys = 4

▽ # lys = 8

□ # lys = 12

a. What happens to the pH of maximum stability as the number of lysine residues increases? Why?

Chapter 7   Experiment 7.4

b. What happens to ΔG_T as the number of lysine residues increases and pH < pK of the lysine side-chain?

c. What happens to ΔG_T as the number of lysine residues increases and pH > pK of the lysine side-chain?

3. Complete and graph the table below to explore the effect of increasing numbers of aspartate residues on ΔG_T.

Values of ΔG_T vs. pH for Different Numbers of Aspartate Residues

# asp	2	3	4	5	6	pH 7	8	9	10	11	12
8	—	—	—	—	—	—	—	—	—	—	—
12	—	—	—	—	—	—	—	—	—	—	—
16	—	—	—	—	—	—	—	—	—	—	—

△ # asp = 8
▽ # asp = 12
□ # asp = 16

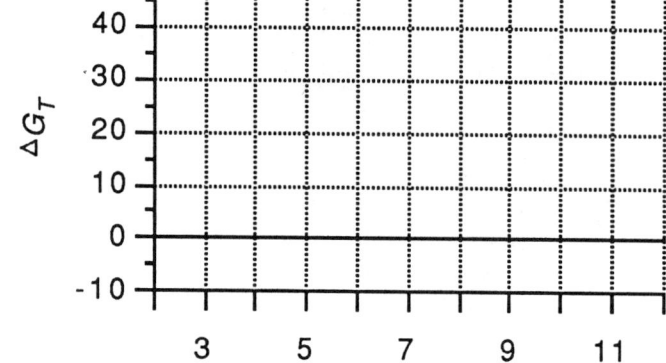

a. What happens to the pH of maximum stability as the number of aspartate residues increases? Why?

b. What happens to ΔG_T as the number of aspartate residues increases and pH < pK of the aspartate side-chain?

c. What happens to ΔG_T as the number of aspartate residues increases and pH > pK of the aspartate side-chain?

pH Dependence of Protein Stability

4. Repeat the calculations in Questions 2 and 3 but plot the fraction of protein in native conformation vs. pH instead of ΔG_T.

Problems

1. The model described in Figure 7.14 does not account for the role that ionic strength plays in modulating the strength of electrostatic interactions. Write a model similar to that described in Figure 7.14 but which includes ionic strength contributions. When ionic strength contributions are considered the energy required to charge a sphere is:

$$\Delta G = \frac{332 q^2}{2 \varepsilon r_p} \left(1 - \frac{0.332 \sqrt{I}\, r_p}{1 + 0.332 \sqrt{I}\, r_p} \right) \qquad (7.19)$$

Explore this model and determine the effect of ionic strength on the protein denaturation curves developed in Questions 1 through 3. Considering only electrostatic effects, will proteins become more or less stable as ionic strength increases?

2. The model in Figure 7.14 predicts that proteins with net charges of $+q$ and $-q$ should be equally stable (Figure 7.15).

This prediction is reasonably correct for some proteins but is very poor for others. Discuss the approximations and simplifications in your model that might be incorrect for some proteins.

FIGURE 7.15

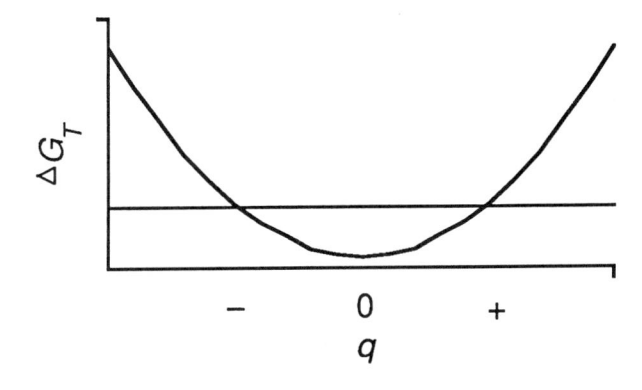

3. Using techniques in molecular biology, it is possible to selectively replace amino acid residues in a protein. Discuss what charged residues you would want in a protein designed to function at high pH. What about a protein that must work at low pH?

Experiment 7.5

Hydrophobic Effects and Macromolecular Structure

Macromolecules exist and function in an aqueous environment. The different chemical groups making up a macromolecule all have differing solubilities in water. Some groups, such as -OH, -CO_2^-, -NH_3^+, are highly soluble in water and are termed *hydrophilic* or *polar*. Other groups, such as alkyl and aromatic groups have low water solubility and are termed *hydrophobic* or *nonpolar*. The three-dimensional conformations of macromolecules are influenced to varying degrees by the differing solubilities of its chemical groups. Although somewhat misleading, the tendency of macromolecules to fold so that the exposure of nonpolar groups to water is minimized is called the *hydrophobic effect*.

The quantitative details of hydrophobic effects on macromolecular structures are not well understood; however, extensive studies of model systems has provided some preliminary conceptual frameworks from which primitive computer simulations may be derived.

The free-energy change associated with partitioning a molecule between two phases can be calculated from the solubilities of the molecule in each phase (e.g., Experiment 3.5). Thus tables of the free energy of transfer of molecules from nonpolar to polar environments can be constructed, and simple empirical equations derived. For example, the free energy of transfer of alkanes between CCl_4 (a nonpolar solvent) and water depends on the number of carbon atoms in the alkane. Every -CH_2- contributes equally to the free energy of transfer:

$$\Delta G_{hw} = 0.884 n_{CH_2} + 2.102 n_{CH_3} \tag{7.20}$$

where ΔG_{hw} = free energy of transfer from CCl_4 to H_2O
n_{CH_2} = no. of methylene groups in the alkane
n_{CH_3} = no. of methyl groups in the alkane

Chapter 7 Experiment 7.5

Table 7.4

Group	G_{hw} (kcal/mol)
-CH$_2$-	0.884
-CH$_3$	2.102
-OH	-2.114
-CO$_2$H	-4.712

Notice that the sign of G_{hw} indicates that -OH and -CO$_2$H (polar groups) prefer the aqueous phase, whereas -CH$_2$- and CH$_3$ (nonpolar groups) prefer the CCl$_4$ phase.

The contents of Table 7.4 may be used to sum the contributions of individual component groups in a molecule and thus calculate the free energy of transfer of the molecule. For example, the free energy of transfer of CH$_3$(CH$_2$)$_6$CH$_3$ from CCl$_4$ to H$_2$O is given by:

$$\Delta G_{hw} = (2 \times 2.102) + (6 \times 0.884) = 9.508 \text{ kcal/mol}$$

The equilibrium constant for the partition of this molecule between the two phases is given by:

$$e^{-\frac{\Delta G_{hw}}{RT}} = 1.31 \times 10^{-7}$$

Directions

Build and verify the model described in Figure 7.16.

Hydrophobic Effects and Macromolecular Structure

Figure 7.16

Formulas

	A	B	C	D
1	Group	Delta Ghw	No.	No.*Delta Ghw
2				
3	-CH2-	0.884	6	=B3*C3
4	-CH3	2.102	2	=B4*C4
5	-OH	-2.114	0	=B5*C5
6	-COOH	-4.712	0	=B6*C6
7				
8			Delta G =	=SUM(D3..D6)
9			Hyd -> Wat	
10				
11			[Wat]/[Hyd] =	=EXP(-D8/0.6)

Values

	A	B	C	D
1	Group	Delta Ghw	No.	No.*Delta Ghw
2				
3	-CH2-	0.884	6	5.304
4	-CH3	2.102	2	4.204
5	-OH	-2.114	0	0
6	-COOH	-4.712	0	0
7				
8			Delta G =	9.508
9			Hyd -> Wat	
10				
11			[Wat]/[Hyd] =	1.31E-7

Questions

1. How many methylene groups are required for the organic acid $CH_3(CH_2)_nCO_2H$ to be more soluble in CCl_4 than in H_2O?

2. How many methylene groups are required for $HO(CH_2)_nCO_2H$ to be more soluble in CCl_4 than in H_2O?

Chapter 7   Experiment 7.5

Problems

1. Membranes are organized as phospholipid bilayers (Chapter 6). The polar head groups are exposed to an aqueous environment, and the hydrocarbon chains of the fatty acids are located in the interior (Figure 7.17). This structural organization is a direct consequence of hydrophobic effects!

    a. Suppose that a molecule of palmitic acid is embedded in a membrane bilayer such that the carboxyl group is adjacent to the polar head groups (Figure 7.18a).

    If the molecule "flips" its orientation such that the carboxyl group becomes buried in the hydrophobic interior and the hydrocarbon chain is exposed to the aqueous environment (Figure 7.18b), then the transfer of one carboxyl group from a polar to a nonpolar environment would have an associated free energy change of +4.712 kcal/mol (the opposite sign as that noted in Table 7.4 because the direction of movement is reversed). The free energy of transfer for each of the 14 methylene groups and one methyl group would be the same as noted in Table 7.4. Redesign the model described in Figure 7.16 to answer the question, What is the free energy of transfer associated with such a flip?

Figure 7.17

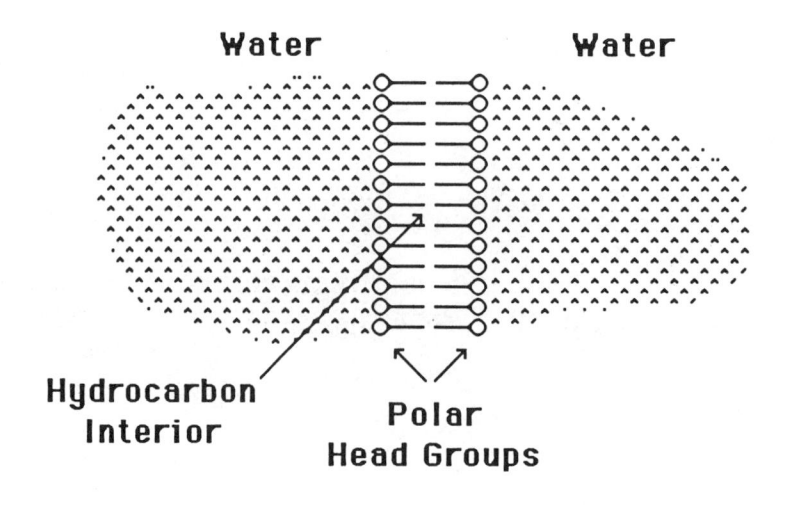

Figure 7.18

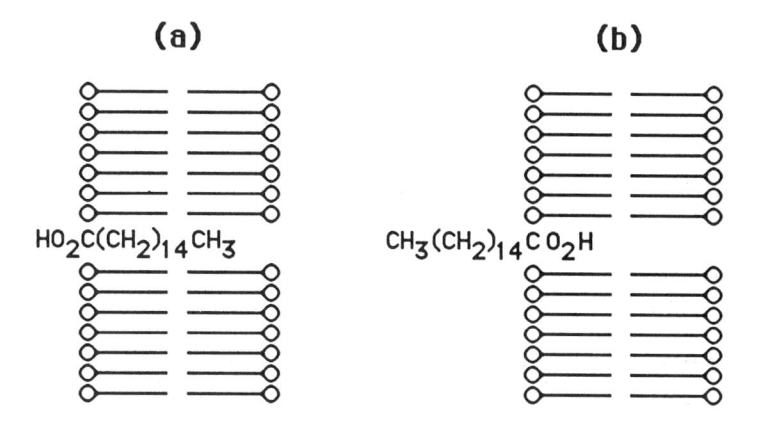

(a) HO₂C(CH₂)₁₄CH₃

(b) CH₃(CH₂)₁₄CO₂H

b. Suppose that the palmitic acid molecule described in Figure 7.18a undergoes a less drastic movement — exposure of the methylene group adjacent to the carboxyl group. Determine the relative fractions of molecules with 0, 1, 2, 3, . . . methylene groups exposed.

These results demonstrate that hydrophobic effects are effective mediators of a smooth membrane surface.

c. What fraction of palmitic acid molecules embedded in a lipid bilayer are completely buried relative to those that have just the carboxyl group exposed?

2. A striking feature of the folded structure of proteins is the tendency for the side chains of nonpolar amino acids to be shielded from the aqueous environment. Unfolded proteins, of course, would expose the side chains of such amino acids to water. A rough estimate of the contribution of hydrophobic effects to the stabilization of the protein can be calculated from the free energies of transfer from ethanol to water as noted in Table 7.5.

Chapter 7 Experiment 7.5

Table 7.5

Amino Acid	G_{hw} (kcal/mol)
Alanine	0.6
Methionine	1.3
Valine	1.5
Leucine	1.7
Tyrosine	2.3
Phenylalanine	2.5
Tryptophan	3.3

Complete the model described in Figure 7.19 and answer the following questions:

a. Ribonuclease has the nonpolar amino acid composition of:

ala	12	leu	2	trp	0
met	4	tyr	6		
val	9	phe	3		

Use the above totals and the model described in Figure 7.19 to calculate the contribution of these residues to the hydrophobic stabilization of the protein.

Because the native conformation of most proteins is only stable by about -10 kcal/mol, the hydrophobic effect represents an essential stabilizing interaction.

b. Comparison of the amino acid sequences of many related proteins reveals that methionine, valine, and leucine can often be interchanged with each other and that tyrosine, phenylalanine, and tryptophan can be interchanged.

Use the model described in Figure 7.19 to examine the effect of changing all valine residues in ribonuclease to either alanine, leucine, or phenylalanine. Do the results support or contradict the conclusion that amino acid substitutions tend to occur so as to conserve the hydrophobic contribution to the free energy of stabilization?

Figure 7.19

Formulas

	A	B	C	D
1	Amino Acid	Delta G	No.	No.*Delta G
2				
3	ala	0.6	12	=B3*C3
4	met	1.3	4	=B4*C4
5	val	1.5	9	=B5*C5
6	leu	1.7	2	=B6*C6
7	tyr	2.3	6	=B7*C7
8	phe	2.5	3	=B8*C8
9	trp	3.3	0	=B9*C9
10				
11			Delta G =	=SUM(D3..D9)
12			Hyd -> Wat	

Values

	A	B	C	D
1	Amino Acid	Delta G	No.	No.*Delta G
2				
3	ala	0.6	12	7.2
4	met	1.3	4	5.2
5	val	1.5	9	13.5
6	leu	1.7	2	3.4
7	tyr	2.3	6	13.8
8	phe	2.5	3	7.5
9	trp	3.3	0	0
10				
11			Delta G =	50.6
12			Hyd -> Wat	

Chapter 7   Experiment 7.5

Conclusion

In closing this chapter you should be aware of two issues:

1. The present understanding of the physical basis of the hydrophobic effect

2. The role of other forces (not covered in this chapter) in stabilizing three-dimensional macromolecular conformations

With respect to the first point, the precise physical basis for hydrophobic effects is not quantitatively understood but appears to be primarily an entropic effect. Water molecules participate in a continually changing pattern of hydrogen bonds with their neighbors. The presence of hydrophobic groups that are unable to participate in these hydrogen bond interactions restricts the number of possible orientations that the water molecule may adopt. This restriction in orientation yields an unfavorable entropic contribution to the overall free energy of the process — an unfavorable contribution that is decreased if the degree of interaction between nonpolar groups and water is minimized by burying hydrophobic groups in the interior of lipid bilayers and proteins.

The precise magnitude of this effect is difficult to estimate. The experiments in this chapter are based on the behavior of model systems that may or may not have quantitative validity in the very different environment surrounding a macromolecule. Consequently, although the models presented in this chapter illustrate qualitative aspects of hydrophobic effects, the quantitative precision of the models are only approximate.

With respect to the second point, two important factors that help determine macromolecular configuration but have not been covered in this chapter are hydrogen bonds and conformational entropy.

Hydrogen bonds are generated by the attractive interaction between lone pair electrons on nitrogen or oxygen atoms and the hydrogen atoms of amino, amide, hydroxyl, or sulfhydryl groups. The properties of water are strongly influenced by hydrogen bond formation, as are the major categories of secondary structure in proteins and nucleic acids (e.g., alpha and double-helical structures).

Because of the nature of hydrogen bond interactions, hydrogen bonds are highly directional and require specific orientations of both donor and acceptor groups. Hydrogen bonds, therefore, provide the thread that helps stitch together the residues in a macromolecule into a specific conformation.

The net contribution of hydrogen bonds to macromolecular stability, however, is difficult to quantitate. Polar groups that are hydrogen bonded to one another in a folded structure are most likely hydrogen bonded to water in an unfolded structure. The net change in free energy consequent to making these trades should be relatively small. It seems likely therefore that because other forces also drive the macromolecule to fold, hydrogen bonding contributes to the energetics of stability by choosing between alternate folded structures — not by choosing between folded and unfolded structures.

Finally, with all the different types of interactions that stabilize macromolecular structure, it may seem surprising that such molecules are only marginally stable and are always on the verge of being denatured. *The major driving force behind denaturation is the increase in conformational entropy that occurs when a macromolecule unfolds.*

The unfolded state should approximate a random coil and have numerous accessible conformations. The number of accessible conformations in folded structures is greatly reduced and is therefore entropically very unfavorable.

Recall the basic definition for free energy:

$$\Delta G = \Delta H - T \Delta S \qquad (7.21)$$

where ΔG = change in free energy upon folding
ΔH = change in enthalpy
T = absolute temperature
ΔS = change in entropy

Because the folding process is entropically unfavorable, $\Delta S < 0$ and the term $-T\Delta S$ will be positive (unfavorable). Furthermore, the contribution of configurational entropy to the overall change in the free-energy of folding should become more positive with increasing temperature. At some temperature, configurational entropy terms exceed stabilizing terms in the overall free-energy equation, and the macromolecule unfolds.

You might enjoy exploring this last consideration by using the simple model described in Figure 7.20. Assume that the enthalpy and entropy changes (ΔH and ΔS, respectively) that result from a change from an unfolded to a folded conformation are:

Chapter 7   Experiment 7.5

$$\Delta H = -10 \text{ kcal/mol/deg}$$
$$\Delta S = -30 \text{ cal/deg/mol}$$

What is the temperature range over which the folded state is the more stable form ($\Delta G < 0$)? The unfolded state? (Assuming that ΔH and ΔS are themselves temperature-independent — an assumption that is, in general, not true.)

This simplified example illustrates that although proteins are stabilized by many interactions, the destabilizing interactions are of comparable magnitude. The overall stability of macromolecules represents the net balance between these opposing forces. As a result, the conformations of macromolecules are quite sensitive to environmental conditions. This sensitivity has important consequences for macromolecule structure and function.

Figure 7.20

Formulas

	A	B	C
1	$\Delta H =$	-10	kcal/mole/deg
2	$\Delta S =$	-30	cal/deg/mole
3	T (°C) =	20	°C
4			
5	ΔG (unfolded to folded) =	=B1-((273+B3)*B2/1000)	

Values

	A	B	C
1	$\Delta H =$	-10	kcal/mole/deg
2	$\Delta S =$	-30	cal/deg/mole
3	T (°C) =	20	°C
4			
5	ΔG (unfolded to folded) =	-1.210	kcal/mole

References

Excellent discussions of macromolecular structure and stability may be found in the following references:

Cantor, C. R. and Schimmel, P. R. *Biophysical Chemistry*, Vols. I, II, III. San Francisco: W. H. Freeman, 1980.

Schultz, G. E. and Schirmer, R. H. *Principles of Protein Structure*. New York: Springer-Verlag, 1979.

Creighton, T. E. *Proteins*. San Francisco: W. H. Freeman, 1983.

Wood, W. B., Wilson, J. H., Benbow, R. M. and Hood, L. E. *Biochemistry: A Problems Approach*, 2nd Edition. Menlo Park: Benjamin/Cummings, 1981.

Appendices

Appendix A:
Common Electronic Spreadsheet Formats

Cell Addresses

Most electronic spreadsheet programs identify columns with letters and rows with numbers. Thus cell names are typically of the form: A1, G3, Z256, etc.

Multiplan (a Microsoft product) uses a different convention. Multiplan identifies both columns and rows with numbers. Cell names are of the form R1C2 (for row 1, column 2). If you are an experienced Multiplan user you should have no trouble translating the models in this book into a Multiplan format, especially if you are used to building formulas by "pointing" rather than typing. (See your owner's manual for an explanation of "pointing").

Distinguishing labels, values, and formulas

Methods for distinguishing labels, values, and formulas vary considerably from one program to the next. Early electronic spreadsheets distinguished cells exclusively by the first character entered into a cell. Recent programs use quite sophisticated techniques and examine the entire entry.

To enter a label first try to simply type it into the cell. Most simple spreadsheet algorithms identify labels by the fact that they usually begin with a letter. If you get an error message or other strange behavior, precede your label with a double-quote. The double-quote will probably be invisible in the cell and will act only as a hidden, label-identifying character. (This latter technique is also useful for entering labels that begin with a number.)

To enter a value type it into the cell. Most simple spreadsheet algorithms identify values by the fact that they begin with a number or with an operator such as + or -.

To enter a formula may require some experimentation (or, you could read the owner's manual). The favored technique of recently designed electronic spreadsheet programs is to begin all formulas with an equal sign, =. Older electronic spreadsheet programs treated formulas identically to values. 37+A3 would be a formula because it begins with a number. (Note, however, that A3+37 might easily be misinterpreted as a label and might require an entry such as +A3+37 instead).

Appendix B
Built-in Formulas

Built-in formulas vary slightly in their syntax from program to program. Visicalc and Lotus 1-2-3, for example, require an @ sign as the first character of a formula. Jazz or Supercalc would be confused by such a symbol. Below are lists of the built-in formulas that are used in this book and their representation in several popular electronic spreadsheet programs.

This Book
⇓

Jazz	Lotus 1-2-3	Visicalc	Supercalc
ABS(B1)	@ABS(B1)	@ABS(B1)	ABS(B1)
COS(B1)	@COS(B1)	@COS(B1)	COS(B1)
EXP(B1)	@EXP(B1)	@EXP(B1)	EXP(B1)
IF(A1=0,B1,B2)	@IF(A1=0,B1,B2)	@IF(A1=0,B1,B2)	IF(A1=0,B1,B2)
INT(B1)	@INT(B1)	@INT(B1)	INT(B1)
LN(B1)	@LN(B1)	@LOG(B1)	LOG(B1)
LOG(B1)	@LOG(B1)	@LOG10(B1)	LOG10(B1)
MAX(B1..B10)	@MAX(B1..B10)	@MAX(B1..B10)	MAX(B1..B10)
MIN(B1..B10)	@MIN(B1..B10)	@MIN(B1..B10)	MIN(B1..B10)
SIN(B1)	@SIN(B1)	@SIN(B1)	SIN(B1)
SQRT(B1)	@SQRT(B1)	@SQRT(B1)	SQRT(B1)
SUM(B1..B10)	@SUM(B1..B10)	@SUM(B1..B10)	SUM(B1..B10)
TAN(B1)	@TAN(B1)	@TAN(B1)	TAN(B1)

Answer Key

Experiment 1.1

1. 41%
2. 46%
3. 29%

Experiment 2.5

1. Cell C1 updates but cell B2 retains its old value.
   Cell B1 updates only on a second round or recalculation.

2. The model has no values at all. When the formula change is made the values depend on the status at the time of the change. Each round of recalculation, however, "updates" the results because the model chases its own tail forever.

3. On most electronic spreadsheet programs cell A1 displays a 1 when the formula is entered. Each round of recalculation increments the value by 1.

   For some reason undecipherable to us, a few spreadsheets give other results.

Experiment 3.1

1.

$[Zap]_o$	$x/[A]_o$	$[B]_o$	$x/[A]_o$
1.00	0.0909	1.00	0.0909
25.00	0.0155	25.00	0.3881
50.00	0.0100	50.00	0.4983
75.00	0.0076	75.00	0.5674
100.00	0.0062	100.00	0.6170

2.

			K_{eq}		
$[B]_o$	0.0001	0.001	0.01	0.1
1	0.01	0.03	0.09	0.24
25	0.05	0.15	0.39	0.76
50	0.07	0.20	0.50	0.85
75	0.08	0.24	0.57	0.89
100	0.10	0.27	0.62	0.92

At high values of K_{eq} the magnitude of the effect of increases in $[B]_o$ is less (e.g., as $[B]_o$ increases 100-fold in the table above, the fold increase in $x/[A]_o$ is reduced by increases in the value of K_{eq}).

3.

$[B]_o$	$[D]_o$	$x/[A]$
50.00	1.00	0.0823
20.00	1.00	0.0487
1.00	1.00	0.0056
1.00	20.00	0.0005
1.00	50.00	0.0002

To decrease the conversion of Zap to Superzap, decrease the ratio $[B]_o:[D]_o$.

Experiment 3.2

1.

pK	pH	[B]/[HB]	Conj. Base	Conj. Acid
4	2	1.E-2	0.0099	0.9901
4	4	1.E0	0.5000	0.5000
4	6	1.E2	0.9901	0.0099
4	8	1.E4	0.9999	0.0001
4	10	1.E6	1.0000	0.0000
4	12	1.E8	1.0000	0.0000
8	2	1.E-6	0.0000	1.0000
8	4	1.E-4	0.0001	0.9999
8	6	1.E-2	0.0099	0.9901
8	8	1.E0	0.5000	0.5000
8	10	1.E2	0.9901	0.0099
8	12	1.E4	0.9999	0.0001

When pH < pK, [B-]/[HB] < 1.
When pH > pK, [B-]/[HB] > 1.
When pH = pK, [B-]/[HB] = 1.

2.

pK	pH	[B]/[HB]	Conj. Base	Conj. Acid
2.3	0	3.E-3	0.0050	0.9950
2.3	2	5.E-1	0.3339	0.6661
2.3	4	5.E1	0.9804	0.0196
2.3	6	5.E3	0.9998	0.0002
2.3	8	5.E5	1.0000	0.0000
2.3	10	5.E7	1.0000	0.0000
2.3	12	5.E9	1.0000	0.0000
2.3	14	5.E11	1.0000	0.0000
9.1	0	8.E-10	0.0000	1.0000
9.1	2	8.E-8	0.0000	1.0000
9.1	4	8.E-6	0.0000	1.0000
9.1	6	8.E-4	0.0008	0.9992
9.1	8	8.E-2	0.0736	0.9264
9.1	10	8.E0	0.8882	0.1118
9.1	12	8.E2	0.9987	0.0013
9.1	14	8.E4	1.0000	0.0000
12.5	0	3.E-13	0.0000	1.0000
12.5	2	3.E-11	0.0000	1.0000
12.5	4	3.E-9	0.0000	1.0000
12.5	6	3.E-7	0.0000	1.0000
12.5	8	3.E-5	0.0000	1.0000
12.5	10	3.E-3	0.0032	0.9968
12.5	12	3.E-1	0.2403	0.7597
12.5	14	3.E1	0.9693	0.0307

Experiment 3.3

1.
pH	[H_3Arg^{++}]	[H_2Arg^+]	[HArg]	[Arg^-]
1	0.95	0.05	0.00	0.00
2	0.67	0.33	0.00	0.00
3	0.17	0.83	0.00	0.00
4	0.02	0.98	0.00	0.00
5	0.00	1.00	0.00	0.00
6	0.00	1.00	0.00	0.00
7	0.00	1.00	0.00	0.00
8	0.00	0.98	0.02	0.00
9	0.00	0.80	0.20	0.00
10	0.00	0.28	0.71	0.00
11	0.00	0.04	0.93	0.03
12	0.00	0.00	0.76	0.24
13	0.00	0.00	0.24	0.76
14	0.00	0.00	0.03	0.97

2. The isoelectric point of arginine equals pH 10.8.
   The predominant ionic form at arginine's isoelectric point is HArg.
   The predominant ionic form at pH 7.2 is H_2Arg.

Experiment 3.4

1.
[B]	[HB]	pH
90	10	7.75
80	20	7.40
70	30	7.17
60	40	6.98
50	50	6.80
40	60	6.62
30	70	6.43
20	80	6.20
10	90	5.85

2.
[B]	[HB]	$\Delta pH/\Delta[B]$
90	10	
80	20	0.0352
70	30	0.0234
60	40	0.0192
50	50	0.0176
40	60	0.0176
30	70	0.0192
20	80	0.0234
10	90	0.0352

3.
[B]	[HB]	pH	$\Delta pH/\Delta[B]$
40	60	6.62	
45	55	6.71	0.0178
41	59	6.64	0.0180
40.1	59.9	6.63	0.0181
40.01	59.99	6.62	0.0181

Thus the instantaneous rate of change of pH with respect to [B^-] is 0.0181.

4. 0.0207

Experiment 3.5

1. pK = 4

pH	[HB]$_W$	[HB]$_O$	[B]$_W$	[B]$_O$
3	0.9891	98.9120	0.0989	0.0000
5	0.9009	90.0893	9.0089	0.0009
7	0.0908	9.0818	90.8183	0.0091
9	0.0010	0.0999	99.8891	0.0100
11	0.0000	0.0010	99.9890	0.0100

pK = 7

pH	[HB]$_W$	[HB]$_O$	[B]$_W$	[B]$_O$
3	0.9901	99.0098	0.0001	0.0000
5	0.9900	99.0001	0.0099	0.0000
7	0.9804	98.0391	0.9804	0.0001
9	0.4975	49.7488	49.7488	0.0050
11	0.0099	0.9899	98.9903	0.0099

pK = 10

pH	[HB]$_W$	[HB]$_O$	[B]$_W$	[B]$_O$
3	0.9901	99.0099	0.0000	0.0000
5	0.9901	99.0099	0.0000	0.0000
7	0.9901	99.0089	0.0010	0.0000
9	0.9891	98.9120	0.0989	0.0000
11	0.9009	90.0893	9.0089	0.0009

3. If you don't intuitively see the answer to this question immediately, look at the figures in the table for pH = 7. When pH = pK, the concentrations of the conjugate acid and conjugate base are equal in the aqueous phase, but the distribution ratios for the two forms are very different, so the ratio in the organic phase is not equal to that in the aqueous phase. Note, by the way, that because of the different partition ratios the total amount of base and the total amount of acid in the system as a whole are far from equal when pH = pK.

4. The added solute distributes between the two phases to bring the partition ratios back to the same values as before the addition. In more chemical terms, the solute moves to the organic phase because, after the addition, the chemical potential of solute in the organic phase is smaller than that in the aqueous phase, even though the actual concentration is larger.

Experiment 3.6

1.

Urine pH	Total Blood Ammonia	Total Urine Ammonia	Ratio Urine/Blood Ammonia
7.4	40	40	1
7.2	40	63	2
6.8	40	158	4
6.4	40	395	10
6.0	40	991	25
5.6	40	2489	62
5.2	40	6252	156
4.8	40	15703	393
4.4	40	39443	986
4.0	40	99077	2477
3.6	40	248868	6222
3.2	40	625128	15628

2. The urine/blood total ammonia ratio would be less than 1.

Experiment 4.1

1.

E_a1	E_a2	k_2/k_1
15.0	15.0	1.00
15.0	14.5	2.32
15.0	14.0	5.40
15.0	13.5	12.50
15.0	13.0	29.20
15.0	12.5	67.70

a. You can obtain the answers to these specific questions by trial and error by varying E_a2 in the left-hand model until the desired ratio is displayed in cell B8. The difference between E1 and E_a2 is the answer to the question.

Note more generally that a change of 1 kcal/mol in the value of E_a changes the velocity by a factor of 5.4, so that a change of 5 kcal/mol causes the velocity to change by a factor of $(5.4)^5$, or about 4600-fold, and a change of 10 kcal/mol changes velocity by a factor of about 21×10^6.

2.

E_a	T = 10-20°C	T = 40-50°C	T = 80-90°C
0	1.00	1.00	1.00
2	1.13	1.10	1.08
4	1.27	1.22	1.17
6	1.44	1.35	1.27
8	1.62	1.49	1.37
10	1.83	1.64	1.48
12	2.07	1.82	1.60
14	2.34	2.01	1.73
16	2.64	2.22	1.87
18	2.98	2.45	2.03
20	3.36	2.70	2.19

a. Smaller

Experiment 4.2

1.

	A	B	C	D	E	F	G
$k_1 =$	1	1	0.01	1	0.01	0.01	1
$k_2 =$	1	1	0.01	1	0.01	0.01	1
$k_3 =$	0.01	1	1	1	1	0.01	0.01
$k_4 =$	0.01	1	1	1	1	0.01	0.01
$k_5 =$	1	0.01	1	1	0.01	1	0.01
$K_s = k_2/k_1 =$	1.00	1.00	1.00	1.00	1.00	1.00	1.00
$K_m = (k_2+k_3)/k_1 =$	1.01	2.00	101.00	2.00	101.00	2.00	1.01
$K =$	1.00	0.51	34.00	1.00	1.00	1.97	0.67

Experiment 4.3

1.

[S]	v/Vmax	v/(Vmax-v)	Reaction Order
0.00	0.00	0.00	
0.05	0.05	0.05	0.95
0.10	0.09	0.10	0.91
0.20	0.17	0.20	0.83
0.50	0.33	0.50	0.67
1.00	0.50	1.00	0.50
2.00	0.67	2.00	0.33
5.00	0.83	5.00	0.17
10.00	0.91	10.00	0.09

c. The order decreases because of increasing saturation of the catalytic sites, as discussed in the text.

d. The momentary order is equal to 1 - v/V_{max}. If you wish to derive that relationship, begin by replacing v by $(V_{max}[S])/(K_m + k[S])$ in the expression for the slope, $d(\ln v)/d(\ln S)$. Differentiation yields the result $1 - [S]/(K_m + [S])$. The second term of that answer is equal to v/V_{max}.

2.

K_s	[S]	$[S]/K_s$	v/V_{max}
0.5	1.25	2.5	0.71
1.0	2.5	2.5	0.71
2.0	5	2.5	0.71
0.5	2.5	5	0.83
1.0	5	5	0.83
2.0	10	5	0.83

3.

K_s	v/V_{max}
0.00	1.00
0.05	0.98
0.10	0.96
0.20	0.93
0.50	0.83
1.00	0.71
2.00	0.56
5.00	0.33
10.00	0.20

Experiment 4.4

1.

k_3	K_m	[S]	k_3/K_s	v	$v/(k_3/K_s)$
0.01	0.01	0.01	1.00E0	5.00E-3	5.00E-3
0.01	0.03	0.01	3.33E-1	2.50E-3	7.50E-3
0.01	0.10	0.01	1.00E-1	9.09E-4	9.09E-3
0.01	0.30	0.01	3.33E-2	3.23E-4	9.68E-3
0.01	1.00	0.01	1.00E-2	9.90E-5	9.90E-3
0.01	3.00	0.01	3.33E-3	3.32E-5	9.97E-3
0.01	10.00	0.01	1.00E-3	9.99E-6	9.99E-3
0.01	30.00	0.01	3.33E-4	3.33E-6	1.00E-2
0.01	100.00	0.01	1.00E-4	1.00E-6	1.00E-2
0.01	0.01	0.1	1.00E0	9.09E-3	9.09E-3
0.01	0.03	0.1	3.33E-1	7.69E-3	2.31E-2
0.01	0.10	0.1	1.00E-1	5.00E-3	5.00E-2
0.01	0.30	0.1	3.33E-2	2.50E-3	7.50E-2
0.01	1.00	0.1	1.00E-2	9.09E-4	9.09E-2
0.01	3.00	0.1	3.33E-3	3.23E-4	9.68E-2
0.01	10.00	0.1	1.00E-3	9.90E-5	9.90E-2
0.01	30.00	0.1	3.33E-4	3.32E-5	9.97E-2
0.01	100.00	0.1	1.00E-4	9.99E-6	9.99E-2
0.01	0.01	1	1.00E0	9.90E-3	9.90E-3
0.01	0.03	1	3.33E-1	9.71E-3	2.91E-2
0.01	0.10	1	1.00E-1	9.09E-3	9.09E-2
0.01	0.30	1	3.33E-2	7.69E-3	2.31E-1
0.01	1.00	1	1.00E-2	5.00E-3	5.00E-1
0.01	3.00	1	3.33E-3	2.50E-3	7.50E-1
0.01	10.00	1	1.00E-3	9.09E-4	9.09E-1
0.01	30.00	1	3.33E-4	3.23E-4	9.68E-1
0.01	100.00	1	1.00E-4	9.90E-5	9.90E-1

a. When [S] = 0.01, $v/(k_3/K_m)$ varies by less than 1% when K_m is between 1 and 100
   When [S] = 0.1, $v/(k_3/K_m)$ varies by less than 1% when K_m is between 10 and 100
   When [S] = 1, $v/(k_3/K_m)$ varies by less than 1% only when K_m is near a value of 100
b. A better measure

2.

k_3	K_s	[S]	k_3/K_s	v	$v/(k_3/K_s)$
0.01	0.01	0.01	1.00E0	5.00E-3	5.00E-3
0.03	0.01	0.01	3.00E0	1.50E-2	5.00E-3
0.10	0.01	0.01	1.00E1	5.00E-2	5.00E-3
0.30	0.01	0.01	3.00E1	1.50E-1	5.00E-3
1.00	0.01	0.01	1.00E2	5.00E-1	5.00E-3
0.01	1.00	0.01	1.00E-2	9.90E-5	9.90E-3
0.03	1.00	0.01	3.00E-2	2.97E-4	9.90E-3
0.10	1.00	0.01	1.00E-1	9.90E-4	9.90E-3
0.30	1.00	0.01	3.00E-1	2.97E-3	9.90E-3
1.00	1.00	0.01	1.00E0	9.90E-3	9.90E-3
0.01	0.01	1.00	1.00E0	9.90E-3	9.90E-3
0.03	0.01	1.00	3.00E0	2.97E-2	9.90E-3
0.10	0.01	1.00	1.00E1	9.90E-2	9.90E-3
0.30	0.01	1.00	3.00E1	2.97E-1	9.90E-3
1.00	0.01	1.00	1.00E2	9.90E-1	9.90E-3

a. All conditions

3. In the simple Michaelis treatment, $v = k_3[ES]$. Thus the rate must always be proportional to k_3. $[S]/K_m = [ES]/[E]$, so that at low concentrations of substrate [ES] is nearly proportional to K_m. At high values of $[S]/K_m$, $[ES]/[E]$ can continue to increase without limit, but the absolute value of [ES] cannot exceed $[E]_{tot}$. Reaction velocity becomes essentially independent of the value of $[S]/K_m$, and thus cannot correlate with $k_3 K_m$. This is merely another statement of the familar fact that the rate of reaction becomes independent of substrate concentration (and so independent of $[S]/K_m$) as the active site approaches saturation with substrate (Experiment 4.5).

Experiment 4.5

1.

	$K_a = 1$, $k_3 = 1$	$K_b = 1$, $k_4 = 1$		$K_a = 5$, $k_3 = 5$	$K_b = 0.2$, $k_4 = 0.2$	
[B]	v(fwd)	v(rvs)	v(net)	v(fwd)	v(rvs)	v(net)
0.0	0.67	0.00	0.67	1.43	0.00	1.43
0.1	0.65	0.03	0.61	1.05	0.05	1.00
0.2	0.62	0.06	0.56	0.83	0.08	0.75
0.5	0.57	0.14	0.43	0.51	0.13	0.38
1.0	0.50	0.25	0.25	0.31	0.16	0.16
2.0	0.40	0.40	0.00	0.18	0.18	0.00
5.0	0.25	0.62	-0.38	0.08	0.19	-0.11
10.0	0.15	0.77	-0.62	0.04	0.19	-0.16

a. v(fwd) decreases in value. This decrease would not occur in a homogeneous reaction because the forward and reverse reactions do not share a limited number of catalytic sites.

2.

[I]	v(fwd)	v(rvs)	v(net)	% Inhib.
0.0	0.33	0.33	0.00	0.00
0.1	0.32	0.32	0.00	3.23
0.3	0.30	0.30	0.00	9.09
1.0	0.25	0.25	0.00	25.00
3.0	0.17	0.17	0.00	50.00
10.0	0.08	0.08	0.00	76.92

a. v(net) = 0

b. The inhibitor becomes less effective against the forward reaction relative to the reverse reaction.

c. The ability of the inhibitor to inhibit increases.

3.

		$K_a = 0.1$ $K_b = 10$ $K_i = 1.0$			$K_a = 0.1$ $K_b = 10$ $K_i = 0.1$		
[A]	[B]	v(fwd)	v(rvs)	v(net)	v(fwd)	v(rvs)	v(net)
10.0	0	4.72		4.72	33.11		33.11
8.0	2	5.80	5.80	5.80	38.11	38.11	38.11
6.0	4	7.53	7.53	7.53	44.88	44.88	44.88
5.0	5	8.85	8.85	8.85	49.26	49.26	49.26
4.0	6	10.73	10.73	10.73	54.59	54.59	54.59
2.0	8	18.66	18.66	18.66	69.64	69.64	69.64
0.0	10		71.43	71.43		96.15	96.15

a. The affinity of substrate and product for the enzyme are different. The two reactions would be inhibited to the same extent if the binding affinities of substrate and product were equal.

b. The rate of the forward reaction is proportional to [EA], and the rate of the reverse reaction is proportional to [EB].
$$[EA]/[E]_{tot} = [A]/K_a/(1+[A]/K_a+[B]/K_b+[I]/K_i)$$
$$[EB]/[E]_{tot} = [B]/K_b/(1+[A]/K_a+[B]/K_b+[I]/K_i)$$
In the absence of inhibitor, the last term of the denominator is zero. When inhibitor is added, the increase in the numerical value of the common denominator causes the same percentage decrease in the values of [EA] and [EB], because [A] and [B] do not change.

c. The ability of the inhibitor to inhibit increases.

Experiment 4.6

1.

pH	pK(B1)	pK(B2)	Reletive Velocity
7	7	5	0.4950
7	7	6	0.4545
7	7	7	0.2500
7	7	8	0.0455
7	7	9	0.0050
7	5	7	0.0050
7	6	7	0.0455
7	7	7	0.2500
7	8	7	0.4545
7	9	7	0.4950
7	8	6	0.8264
7	6	8	0.0083

a. v increases as the pK of B1 increases. v decreases as the pK of B2 increases.

b. Examine your plots of the tables above and notice that reaction rate increases as the proton donor (B1) becomes a weaker acid and reaction rate increases as the proton acceptor (B2) becomes a stronger acid.

For a given pH, the weaker acid is more likely to possess a donatable proton and the stronger acid more likely to be able to accept a proton.

2 b. The pH-response curve is sharp when the pKa values of B1 and B2 are equal and becomes broader as the pKa values diverge. As in Question 1, the maximal reaction velocity is greater when the proton-donating group is the weaker acid.

If the pK values are interchanged, for example, from pK values of 5 and 9 for B1 and B2 to 9 and 5, the maximal velocity changes by a large factor, but the shape of the curve does not change. (You cannot really see that result clearly from your results so far; the model of Figure 4.18 is designed to demonstrate such relationships. See Problem 3.)

Experiment 4.7

1.

[S]	$n=1$ S(0.5) = 1	$n=2$ S(0.5) = 1	$n=4$ S(0.5) = 1
	Reaction	Rate	
1.E-8	0.0000	0.0000	0.0000
0.01	0.0099	0.0001	0.0000
0.02	0.0196	0.0004	0.0000
0.05	0.0476	0.0025	0.0000
0.1	0.0909	0.0099	0.0001
0.2	0.1667	0.0385	0.0016
0.5	0.3333	0.2000	0.0588
1	0.5000	0.5000	0.5000
2	0.6667	0.8000	0.9412
5	0.8333	0.9615	0.9984
10	0.9091	0.9901	0.9999
20	0.9524	0.9975	1.0000
50	0.9804	0.9996	1.0000

2 a.

[S]	$n=1$ S(0.5) = 0.1	$n=1$ S(0.5) = 0.2	$n=1$ S(0.5) = 0.4
	Reaction	Rate	
1.E-8	0.0000	0.0000	0.0000
0.01	0.0909	0.0476	0.0244
0.02	0.1667	0.0909	0.0476
0.05	0.3333	0.2000	0.1111
0.1	0.5000	0.3333	0.2000
0.2	0.6667	0.5000	0.3333
0.5	0.8333	0.7143	0.5556
1	0.9091	0.8333	0.7143

Low values of [S] cause the greatest change in reaction velocity with respect to S(0.5).

A modifier cannot be an effective regulator of reaction rates when [S] >> S(0.5).

b.

[S]	$n=2$ S(0.5) = 0.1	$n=2$ S(0.5) = 0.2	$n=2$ S(0.5) = 0.4
	Reaction	Rate	
1.E-8	0.0000	0.0000	0.0000
0.01	0.0099	0.0025	0.0006
0.02	0.0385	0.0099	0.0025
0.05	0.2000	0.0588	0.0154
0.1	0.5000	0.2000	0.0588
0.2	0.8000	0.5000	0.2000
0.5	0.9615	0.8621	0.6098
1	0.9901	0.9615	0.8621

c.

[S]	$n=4$ S(0.5) = 0.1	$n=4$ S(0.5) = 0.2	$n=4$ S(0.5) = 0.4
	Reaction	Rate	
1.E-8	0.0000	0.0000	0.0000
0.01	0.0001	0.0000	0.0000
0.02	0.0016	0.0001	0.0000
0.05	0.0588	0.0039	0.0002
0.1	0.5000	0.0588	0.0039
0.2	0.9412	0.5000	0.0588
0.5	0.9984	0.9750	0.7094
1	0.9999	0.9984	0.9750

Experiment 4.8

1.

| | [S] = 0.25 | | | [S] = 2.5 | | | [S] = 25 | | |
| | | | (Normalized) | | | (Normalized) | | | (Normalized) |
Frt Bnd	tot vel	K_s eff	K_s eff	tot vel	K_s eff	K_s eff	tot vel	K_s eff	K_s eff
0.0	4.76	5.00	0.00	33.33	5.00	0.00	83.33	5.00	0.00
0.1	9.29	2.44	0.54	39.09	3.90	0.23	84.90	4.45	0.12
0.2	13.81	1.56	0.72	44.85	3.07	0.41	86.47	3.91	0.23
0.3	18.33	1.11	0.82	50.61	2.44	0.54	88.04	3.40	0.34
0.4	22.86	0.84	0.88	56.36	1.94	0.65	89.60	2.90	0.44
0.5	27.38	0.66	0.91	62.12	1.52	0.73	91.17	2.42	0.54
0.6	31.90	0.53	0.94	67.88	1.18	0.80	92.74	1.96	0.64
0.7	36.43	0.44	0.96	73.64	0.90	0.86	94.31	1.51	0.73
0.8	40.95	0.36	0.98	79.39	0.65	0.92	95.87	1.08	0.83
0.9	45.48	0.30	0.99	85.15	0.44	0.96	97.44	0.66	0.91
1.0	50.00	0.25	1.00	90.91	0.25	1.00	99.01	0.25	1.00

a. When [S] = 0.25, [S] is only 5% as large as K_S in the absence of modifier. Thus the velocity is slightly less than 5% of V_{max}. If modifier is added until 10% of the enzyme molecules bind modifier, the K_S of those molecules is changed to 0.25. Because [S] = 0.25, the 10% of enzyme molecules binding modifier catalyzes the reaction at 50% of V_{max}, which is equal to 5% of V_{max} for the whole population of enzyme. Thus the velocity is approximately twice that in the absence of modifier.

Because velocity is proportional to $[S]/(K_S + [S])$, when $K_S \gg [S]$, it would be necessary to decrease K_S by a factor of about two to double the rate if all molecules had the same value of K_S. Thus the effective K_S for the whole population that would yield the same rate of reaction as given by the mixture of 10% of molecules that bear modifier and 90% that do not would be about half the initial value of K_S.

When [S] is large compared to the initial value of K_S, much more modifier must bind before the effective K_S reaches its midpoint. That result, seen in the tables, follows from reasoning similar to that in the preceding paragraph.

b. Each time the fraction of enzyme that binds modifier increases by 10%, for example, the amount of unbound enzyme necessarily decreases by 10%. The rate of reaction catalyzed by enzyme bearing modifier increases by 10%, and the rate of reaction catalyzed by enzyme not bearing modifier decreases by 10%. The difference between those changes is the net increase in velocity. At any given substrate concentration, the rate of reaction catalyzed by each molecule of each form is constant. Thus the increment in velocity is equal for each step in the tables. (Remember that, because modifier binds in a Michaelis fashion, it is necessary to add a larger amount of modifier at each step to increase the fraction of bound enzyme by 10%.)

c. Yes, the incremental velocity must be proportional to the fraction of enzyme that bears modifier for the reasons given in answer b. The dissociation constant of the modifier can therefore be estimated from the type of plot suggested. When the reaction velocity is midway between the velocity in the absence of modifier and the velocity at a saturating level of modifier, half of the enzyme molecules bear modifier, and the concentration of modifier is thus equal to its dissociation constant.

Experiment 4.9

1.

[S]	log[S]	[E]	[ES]	[ES2]	[ES3]	[ES4]	Sites	Michaelis Sites
0		1.00	0.00	0.00	0.00	0.00	0.00	0.00
0.01	-2.00	0.96	0.04	0.00	0.00	0.00	0.01	0.01
0.0316	-1.50	0.88	0.11	0.01	0.00	0.00	0.03	0.03
0.1	-1.00	0.68	0.27	0.04	0.00	0.00	0.09	0.09
0.316	-0.50	0.33	0.42	0.20	0.04	0.00	0.24	0.24
1	0.00	0.06	0.25	0.38	0.25	0.06	0.50	0.50
3.16	0.50	0.00	0.04	0.20	0.42	0.33	0.76	0.76
10	1.00	0.00	0.00	0.04	0.27	0.68	0.91	0.91
31.6	1.50	0.00	0.00	0.01	0.11	0.88	0.97	0.97
100	2.00	0.00	0.00	0.00	0.04	0.96	0.99	0.99

2.

[S]	K	[S]/K	Fracton Sites Filled 4-Site	1-Site
0.25	1	0.25	0.20	0.20
0.50	2	0.25	0.20	0.20
1.00	4	0.25	0.20	0.20
2.00	8	0.25	0.20	0.20
0.25	0.5	0.50	0.33	0.33
0.50	1	0.50	0.33	0.33
1.00	2	0.50	0.33	0.33
2.00	4	0.50	0.33	0.33

3. Yes, it must be true. Think about it!

Experiment 4.10

1.

[S]	log[S]	[E]	[ES]	[ES2]	[ES3]	[ES4]	Frtn Sites Bound
0		1.00	0.00	0.00	0.00	0.00	0.00
0.01	-2.00	0.96	0.04	0.00	0.00	0.00	0.01
0.0316	-1.50	0.88	0.11	0.01	0.00	0.00	0.03
0.1	-1.00	0.68	0.27	0.04	0.00	0.00	0.09
0.316	-0.50	0.33	0.42	0.20	0.04	0.00	0.24
1	0.00	0.06	0.25	0.38	0.25	0.06	0.50
3.16	0.50	0.00	0.04	0.20	0.42	0.33	0.76
10	1.00	0.00	0.00	0.04	0.27	0.68	0.91
31.6	1.50	0.00	0.00	0.01	0.11	0.88	0.97
100	2.00	0.00	0.00	0.00	0.04	0.96	0.99

a. They are, of course, identical.

2.

[S]	log[S]	[E]	[ES]	[ES2]	[ES3]	[ES4]	Frtn Sites Bound
0		1.00	0.00	0.00	0.00	0.00	0.00
0.01	-2.00	1.00	0.00	0.00	0.00	0.00	0.00
0.0316	-1.50	0.99	0.01	0.00	0.00	0.00	0.00
0.1	-1.00	0.96	0.04	0.00	0.00	0.00	0.01
0.316	-0.50	0.84	0.12	0.03	0.01	0.01	0.06
1	0.00	0.31	0.14	0.10	0.14	0.31	0.50
3.16	0.50	0.01	0.01	0.03	0.12	0.84	0.94
10	1.00	0.00	0.00	0.00	0.04	0.96	0.99
31.6	1.50	0.00	0.00	0.00	0.01	0.99	1.00
100	2.00	0.00	0.00	0.00	0.00	1.00	1.00

Experiment 4.11

1. ΔΔG(bind) =

	0		−0.9		−1.4		−2.8	
[S]	Hill	True	Hill	True	Hill	True	Hill	True
0.010	1.00	0.99	1.01	1.01	1.01	1.01	1.01	1.01
0.033	1.00	0.97	1.04	1.04	1.03	1.03	1.09	1.09
0.100	1.00	0.91	1.16	1.15	1.21	1.20	2.50	2.50
0.200	1.00	0.83	1.43	1.39	1.74	1.72	3.65	3.64
0.333	1.00	0.75	1.87	1.75	2.55	2.48	3.91	3.86
0.500	1.00	0.67	2.34	2.01	3.17	2.90	3.97	3.73
1.000	1.00	0.50	2.79	1.39	3.57	1.78	3.99	1.99
1.500	1.00	0.40	2.62	0.65	3.44	0.66	3.98	0.66
2.000	1.00	0.33	2.34	0.32	3.17	0.27	3.97	0.24
3.000	1.00	0.25	1.87	0.12	2.55	0.07	3.91	0.05
5.000	1.00	0.17	1.43	0.04	1.74	0.02	3.65	0.01
10.000	1.00	0.09	1.16	0.01	1.21	0.00	2.50	0.00
100.000	1.00	0.01	1.01	0.00	1.01	0.00	1.01	0.00

c. Hill orders assume their maximum value when [S] = S(0.5)

d. At low values of [S] there is little catalytic site saturation.
At high values of [S] the true order must decrease towards zero at saturation. The Hill equation was devised to mathematically compensate for saturation.

2.

	ΔΔG$_{bind}$ = −1.4		ΔΔG$_{bind}$ = −2.8	
[S]	Hill	True	Hill	True
0.500	3.17	2.90	3.97	3.73
1.000	3.57	1.78	3.99	1.99
2.000	3.17	0.27	3.97	0.24
0.333	2.55	2.48	3.91	3.86
1.000	3.57	1.78	3.99	1.99
3.000	2.55	0.07	3.91	0.05
0.200	1.74	1.72	3.65	3.64
1.000	3.57	1.78	3.99	1.99
5.000	1.74	0.02	3.65	0.01

3. a. As ΔΔG$_{bind}$ increases in magnitude, the maximum value of the Hill order asymptotically approaches the number of catalytic sites.

   b. As ΔΔG$_{bind}$ assumes larger negative values, the rate of change of the Hill order decreases near the midpoint, and a broader range of [S] values exhibits "infinite" cooperativity.

Experiment 4.12

3. Intermediate enzyme-substrate complexes (ES$_1$, ES$_2$, and ES$_3$) exist in lower concentrations for all values of [S] when binding is cooperative.

5. There are no interactions between the sites that bind modifier and the sites that bind substrate, hence no effect.

Experiment 4.13

1 a.

[S]	[I]	% Inhib.
0.001	0.001	1.0
0.01	0.01	9.0
0.1	0.1	47.6
1	1	83.3
10	10	90.1

b.

[S]	[I] = [S]/4	[I] = [S]	[I] = 4*[S]
0.025	0.11	0.10	0.06
0.05	0.21	0.16	0.08
0.1	0.37	0.24	0.10
0.25	0.67	0.33	0.11
0.5	0.91	0.38	0.12
1	1.11	0.42	0.12
2.5	1.28	0.44	0.12
5	1.35	0.45	0.12
10	1.39	0.45	0.12
25	1.41	0.45	0.12
50	1.42	0.45	0.12
100	1.42	0.45	0.12

2.

[S]	[A] = [S]/4	[A] = [S]	[A] = 4*[S]
0.025	0.00	0.00	0.03
0.05	0.01	0.03	0.14
0.1	0.11	0.25	0.20
0.25	0.69	0.49	0.20
0.5	0.79	0.50	0.20
1	0.80	0.50	0.20
2.5	0.80	0.50	0.20
5	0.80	0.50	0.20
10	0.80	0.50	0.20
25	0.80	0.50	0.20
50	0.80	0.50	0.20
100	0.80	0.50	0.20

b. The analog mimics the substrate in causing the affinity for substrate (and analog) at other sites to increase. Thus when most sites are unfilled, increasing the concentration of analog actually causes an increase in the amount of substrate bound. At higher total ligand concentrations, most sites are filled, and an increase in the concentration of analog can only cause it to compete more effectively with substrate and thus decrease the amount of substrate bound.

c. Because the binding curve is steeper, and saturation is approached at lower values of [ligand]/S(0.5).

3.

[Analog]	Reaction Rate − Analog	+ Analog	% Analog Effect
0	0.0250	0.0250	
0.005	0.0250	0.0350	40.3
0.01	0.0250	0.0471	88.5
0.03	0.0250	0.1106	343.2
0.05	0.0250	0.1761	605.4
0.1	0.0250	0.2267	808.2
0.3	0.0250	0.1168	367.8
0.5	0.0250	0.0740	196.4
1	0.0250	0.0385	54.1
3	0.0250	0.0132	−47.3
5	0.0250	0.0079	−68.2
10	0.0250	0.0040	−84.0

a. No

Experiment 5.1

1.

$[A]_o$	$[B]_o$	$[C]_o$	$[D]_o$	$[C]_{eq}/[A]_o$
1	10	0	0	0.27
5	50	0	0	0.27
10	100	0	0	0.27
50	500	0	0	0.27

2 b.

[FDP]$_o$	[3-PGAld]$_{eq}$	[3-PGAld]$_{eq}$/[FDP]$_o$
1	0.0100	0.0101
0.1	0.0031	0.0321
0.01	0.0010	0.1051
0.001	0.0003	0.3702
0.0001	0.0001	1.6180

Experiment 5.2

1.

G°(C)	[A]$_{eq}$	[C]$_{eq}$	[P]$_{eq}$	[P]$_{eq}$/[A]$_{eq}$
-2	0.1131	3.2392	92.8033	820.8033
-3	0.0991	15.1957	81.3358	820.8033
-4	0.0596	48.9562	48.9562	820.8033
-5	0.0191	83.6964	15.6368	820.8033
-6	0.0041	96.4885	3.3679	820.8033
-7	0.0008	99.3247	0.6477	820.8033

a. [P]$_{eq}$ decreases.

b. No

c. The mass accumulates as C.

Experiment 5.3

1.

F	[B]	[C]	[D]	[P]
0.2	98.0	97.0	9.5	45.5
0.4	96.0	94.0	9.0	41.0
0.6	94.0	91.0	8.5	36.5
0.8	92.0	88.0	8.0	32.0
1.0	90.0	85.0	7.5	27.5
1.2	88.0	82.0	7.0	23.0
1.4	86.0	79.0	6.5	18.5
1.6	84.0	76.0	6.0	14.0
1.8	82.0	73.0	5.5	9.5
2.0	80.0	70.0	5.0	5.0

a. Decrease

b. See Question 2 below.

2.

F	k_2[B]	k_{-2}[C]	k_4[D]	k_{-4}[P]
0.2	19.6	19.4	4.8	4.6
0.4	19.2	18.8	4.5	4.1
0.6	18.8	18.2	4.2	3.6
0.8	18.4	17.6	4.0	3.2
1.0	18.0	17.0	3.8	2.8
1.2	17.6	16.4	3.5	2.3
1.4	17.2	15.8	3.2	1.9
1.6	16.8	15.2	3.0	1.4
1.8	16.4	14.6	2.8	1.0
2.0	16.0	14.0	2.5	0.5

a. The concentrations of the intermediates decrease. As flux increases, the reaction rate of B to C decreases but the reaction rate of C to B decreases even more.

b. The situation is the same as in part a. If the reaction D to C decreases more than the reaction C to D, net conversion of C to D will increase.

5. $F = k_3[A]$. This is possible in the nonenzymic case because, no matter how small the later rate constants may be, the concentrations of the intermediates can increase until the rate of each step is as great as the initial step. (In this model, no factors that might limit concentrations in the real world, such as solubility, are taken into account.)

6. Set [P] = 0 in your model. Most of the rate constants contribute to this case. Vary the constants to verify that $k_3[A]$ is the maximal possible rate, but that in this case (unlike Question 5) the reverse rate constants affect the rate even though no full reverse reaction is possible.

7. Two generalizations apply here. (1) Changes in the amount of the first catalyst are likely to have the greatest effect because $k_1[A]$ sets the upper limit of flux. (2) Changing the concentration of a catalyst that is present at higher activity than the others is likely to have very little effect, because such catalysts are not significantly limiting.

Experiment 5.4

1.
Flux	$V_{max} = 100$	$V_{max} = 200$
0	0.00	0.00
20	0.25	0.11
40	0.67	0.25
60	1.50	0.43
80	4.00	0.67
90	9.00	0.82
95	19.00	0.90
99	99.00	0.98
120		1.50
140		2.33
160		4.00
180		9.00
190		19.00
198		99.00

a. When V_{max} = 100, [S] approaches infinity as flux approaches 100.
When V_{max} = 200, [S] approaches K_m as flux approaches 100.

2.
V_{max}	Flux = 100	Flux = 200
1000	0.11	0.25
800	0.14	0.33
600	0.20	0.50
400	0.33	1.00
300	0.50	2.00
220	0.83	10.00
210	0.91	20.00
202	0.98	100.00
200	1.00	
150	2.00	
110	10.00	
105	20.00	
101	100.00	

a. [S] rises catastrophically.

3 a. [S] increases by about 230%.
b. [S] increases by about 20%.

Experiment 5.5

2. An increase in flux through the main sequence will probably cause an increase in the concentration of the branch point metabolite. That would cause the flow through the branch pathway to increase. Sensitivity to control would tend to decrease with increase in the concentration of the branch point metabolite.

TABLE A

V_{max-1}	V_{max-2}	F	S(0.5)	v_2	v_2/V_{max}	Sens. Ind.
1000	100	100	0.1	96.89	0.97	1.58
1000	100	100	0.2	88.19	0.88	4.92
1000	100	100	0.3	75.56	0.76	15.65
1000	100	100	0.5	47.16	0.47	68.90
1000	100	100	0.6	34.94	0.35	105.06
1000	100	100	0.8	17.92	0.18	169.48
1000	100	100	1.0	9.09	0.09	209.65
1000	100	100	1.5	2.10	0.02	244.84

TABLE B

V_{max-1}	V_{max-2}	F	S(0.5)	v_2	v_2/V_{max}	Sens. Ind.
1000	100	10	0.6	4.26	0.04	140.48
1000	100	30	0.6	12.24	0.12	131.87
1000	100	100	0.6	34.94	0.35	105.06
1000	100	300	0.6	69.77	0.70	55.04
1000	100	500	0.6	84.58	0.85	29.23
1000	100	700	0.6	92.28	0.92	14.45

TABLE C

V_{max-1}	V_{max-2}	F	S(0.5)	v_2	v_2/V_{max}	Sens. Ind.
500	500	10	0.6	8.84	0.02	29.65
500	500	30	0.6	26.41	0.05	28.44
500	500	100	0.6	86.76	0.17	24.27
500	500	300	0.6	244.74	0.49	13.11
500	500	500	0.6	367.65	0.74	4.97
500	500	700	0.6	444.74	0.89	2.03

TABLE D

V_{max-1}	V_{max-2}	F	S(0.5)	v_2	v_2/V_{max}	Sens. Ind.
500	500	100	0.1	99.99	0.20	1.03
500	500	100	0.2	99.80	0.20	1.44
500	500	100	0.3	99.00	0.20	2.90
500	500	100	0.5	92.96	0.19	13.21
500	500	100	0.6	86.76	0.17	24.27
500	500	100	0.8	69.16	0.14	59.01
500	500	100	1.0	50.00	0.10	103.12
500	500	100	1.5	18.52	0.04	192.25

TABLE E

V_{max-1}	V_{max-2}	F	S(0.5)	v_2	v_2/V_{max}	Sens. Ind.
1000	10	100	0.1	9.99	1.00	1.25
1000	10	100	0.2	9.84	0.98	4.70
1000	10	100	0.3	9.25	0.92	18.28
1000	10	100	0.5	6.23	0.62	90.46
1000	10	100	0.6	4.49	0.45	134.56
1000	10	100	0.8	2.09	0.21	197.91
1000	10	100	1.0	0.99	0.10	228.23
1000	10	100	1.5	0.21	0.02	249.94

TABLE F

V_{max-1}	V_{max-2}	F	S(0.5)	v_2	v_2/V_{max}	Sens. Ind.
1000	1	100	0.6	0.46	0.46	137.92
1000	3	100	0.6	1.37	0.46	137.17
1000	10	100	0.6	4.49	0.45	134.56
1000	30	100	0.6	12.74	0.42	127.33
1000	100	100	0.6	34.94	0.35	105.06
1000	300	100	0.6	65.24	0.22	65.87
1000	500	100	0.6	76.96	0.15	46.97
1000	700	100	0.6	82.89	0.12	36.39

Experiment 5.6

P-ase V_{max} =	5	4	3	2	1	0.5
B-ase v/V_{max}	0.43	0.36	0.28	0.20	0.11	0.06
Flux	1.71	1.44	1.14	0.82	0.45	0.25
[P]	3.14	3.38	3.69	4.14	4.94	5.83
[D]	0.27	0.22	0.17	0.11	0.06	0.03
Sens. Index*	-2.17	-2.46	-2.77	-3.14	-3.80	-5.32

P-ase K_m =	1	2	3	5	10	15
B-ase v/V_{max}	0.38	0.32	0.28	0.22	0.15	0.12
Flux	1.53	1.28	1.11	0.89	0.62	0.49
[P]	3.29	3.54	3.73	4.02	4.50	4.83
[D]	0.24	0.19	0.16	0.12	0.08	0.06
Sens. Index*	-2.36	-2.65	-2.82	-3.04	-3.32	-3.38

*Sens. Index may vary in your results, depending upon the increment you chose.

2. P is the feedback inhibitor; the system has evolved so that flux will decrease when [P] increases. The concentration of D is dependent on flux; when flux increases, [D] increases until it is high enough to support the new rate of reaction.

3. As you have seen previously, the shapes of enzyme-response curves are such that sensitivity is always greatest at low rates of reaction.

Experiment 5.7

1.

[C]	[D]	K_c(RXN I)	K_d(RXN II)	v(C→D)	v(D→C)	v_{net}	% Waste
1	1	10	0.10	0.91	9.09	-8.18	11.18
1	1	5	0.20	1.67	8.33	-6.66	25.05
1	1	2.5	0.40	2.86	7.14	-4.28	66.71
1	1	1	1.00	5.00	5.00	0.00	
1	1	0.4	2.50	7.14	2.86	4.28	66.73
1	1	0.2	5.00	8.33	1.67	6.66	25.08
1	1	0.1	10.00	9.09	0.92	8.17	11.23
0.3	0.3	1	1.00	2.31	2.31	0.00	
0.3	0.3	0.4	2.50	4.29	1.07	3.21	33.38
0.3	0.3	0.2	5.00	6.00	0.57	5.43	10.48
0.3	0.3	0.1	10.00	7.50	0.30	7.20	4.15
0.1	0.1	1	1.00	0.91	0.91	0.00	
0.1	0.1	0.4	2.50	2.00	0.39	1.61	23.85
0.1	0.1	0.2	5.00	3.33	0.20	3.14	6.31
0.1	0.1	0.1	10.00	5.00	0.10	4.90	2.12

b. High net conversion and low percent waste result when the rates of the two reactions are very different. This is likely to result when the affinities of the enzymes are very different. Of course, it is really [C]/K_c and [D]/K_d, rather than the affinities alone, that is important.

c. The results of increasing the Michaelis constant would be the same as the results of decreasing the substrate concentration.

2.

[C]	[D]	K_c(RXN I)	K_d(RXN II)	v(C→D)	v(D→C)	v_{net}	% Waste
1	1	10	0.10	0.15	9.90	-9.75	1.53
1	1	5	0.20	0.40	9.62	-9.22	4.31
1	1	2.5	0.40	1.38	8.62	-7.24	19.10
1	1	1	1.00	5.00	5.00	0.00	ERR
1	1	0.4	2.50	8.62	1.38	7.24	19.10
1	1	0.2	5.00	9.62	0.40	9.22	4.31
1	1	0.1	10.00	9.90	0.15	9.75	1.53
0.3	0.3	1	1.00	0.83	0.83	0.00	ERR
0.3	0.3	0.4	2.50	3.60	0.14	3.46	4.12
0.3	0.3	0.2	5.00	6.92	0.04	6.89	0.55
0.3	0.3	0.1	10.00	9.00	0.02	8.98	0.19
0.1	0.1	1	1.00	0.10	0.10	0.00	ERR
0.1	0.1	0.4	2.50	0.59	0.02	0.57	2.80
0.1	0.1	0.2	5.00	2.00	0.00	2.00	0.21
0.1	0.1	0.1	10.00	5.00	0.00	5.00	0.04

Experiment 5.8

2 b. The equilibrium constant is more favorable with a stronger oxidizing agent.

Experiment 6.1

1.

Time(s)	Acetamide	Butyramide	Glycerol
0	0	0	0
5	0.9655	0.5530	0.0015
10	0.9988	0.8002	0.0031
15	1.0000	0.9107	0.0046
20	1.0000	0.9601	0.0061
25	1.0000	0.9822	0.0077
100	1.0000	1.0000	0.0303
300	1.0000	1.0000	0.0881
500	1.0000	1.0000	0.1425
700	1.0000	1.0000	0.1936
900	1.0000	1.0000	0.2417
1000	1.0000	1.0000	0.2647
10000	1.0000	1.0000	0.9538

2 a.

V_{cell}	Area	t=200 s	t=2000 s
9.7E-11	1.0E-6	0.04	0.35
9.7E-11	2.0E-6	0.08	0.58
9.7E-11	3.0E-6	0.12	0.73
9.7E-11	4.0E-6	0.16	0.82
9.7E-11	5.0E-6	0.19	0.89
9.7E-11	6.0E-6	0.23	0.93
9.7E-11	7.0E-6	0.26	0.95
9.7E-11	8.0E-6	0.29	0.97
9.7E-11	9.0E-6	0.32	0.98

b.

V_{cell}	Area	$t=200$ s	$t=2000$ s
1.59E-10	1.42E-6	0.0368	0.3128
9.70E-11	1.42E-6	0.0596	0.4593
9.00E-11	1.42E-6	0.0641	0.4845
8.00E-11	1.42E-6	0.0718	0.5255
7.00E-11	1.42E-6	0.0817	0.5734
6.00E-11	1.42E-6	0.0946	0.6299
5.00E-11	1.42E-6	0.1124	0.6966
4.00E-11	1.42E-6	0.1385	0.7749
3.00E-11	1.42E-6	0.1803	0.8630
2.00E-11	1.42E-6	0.2578	0.9493
1.00E-11	1.42E-6	0.4492	0.9974

3.

V_{cell}	Area	$t=200$ s	$t=2000$ s
1.E-10	1.04E-6	0.0428	0.3544
2.E-10	1.65E-6	0.0341	0.2934
3.E-10	2.17E-6	0.0299	0.2617
4.E-10	2.63E-6	0.0272	0.2409
5.E-10	3.05E-6	0.0253	0.2258
6.E-10	3.44E-6	0.0238	0.2140
7.E-10	3.81E-6	0.0226	0.2045
8.E-10	4.17E-6	0.0216	0.1965
9.E-10	4.51E-6	0.0208	0.1897

Experiment 6.2

1.

$[S]_{med}$	$\log[S]_{med}$	$t(1/2)$
1.0E-6	-6	6.44
1.0E-5	-5	6.45
1.0E-4	-4	6.59
1.0E-3	-3	8.02
1.0E-2	-2	22.26
1.2E-2	-1.92	25.43
1.5E-2	-1.82	30.18
1.7E-2	-1.77	33.34
2.0E-2	-1.70	38.09

2.

K_m	$[S]_{med} = 1mM$	$[S]_{med} = 5mM$	$[S]_{med} = 10mM$
0.001	2.82	9.15	17.06
0.002	4.06	10.39	18.30
0.003	5.30	11.63	19.54
0.004	6.53	12.86	20.78
0.005	7.77	14.10	22.02
0.006	9.01	15.34	23.25
0.007	10.25	16.58	24.49
0.008	11.48	17.82	25.73
0.009	12.72	19.05	26.97

Experiment 6.3

1.	[S]$_{med}$	Passive	Facilitated	Combined
	0.00	0.00E0	0.00E0	0.00E0
	0.01	3.07E-6	9.08E-6	1.22E-5
	0.02	6.15E-6	1.09E-5	1.70E-5
	0.03	9.22E-6	1.17E-5	2.09E-5
	0.04	1.23E-5	1.21E-5	2.44E-5
	0.05	1.54E-5	1.24E-5	2.78E-5
	0.06	1.84E-5	1.26E-5	3.10E-5
	0.07	2.15E-5	1.27E-5	3.42E-5
	0.08	2.46E-5	1.28E-5	3.74E-5
	0.09	2.77E-5	1.29E-5	4.06E-5
	0.10	3.07E-5	1.30E-5	4.37E-5

a. Saturation of the carrier site

Experiment 6.4

1.	[S]$_{cell}$	[S]$_{med}$	[S]$_{cell}$/[S]$_{med}$	$\Delta G'$
	0.001	1	0.001	-4.08
	0.01	1	0.01	-2.72
	0.1	1	0.1	-1.36
	1	1	1	0.00
	10	1	10	1.36
	100	1	100	2.72
	1000	1	1000	4.08

Experiment 6.5

1. 236,108

2. 1.32E16

3. 61.78

4.
PEP	glycerol-1-phosphate
7.82E10	41.62
4.79E32	72,091
4,273	3.46

Experiment 6.6

1 a.	$\Delta G^{\circ\prime}$	Charge	$\Delta \Psi_{cell-med}$	[S]$_{cell}$/[S]$_{med}$
	0	1	-0.2	2431.06
	0	1	-0.16	511.25
	0	1	-0.12	107.52
	0	1	-0.08	22.61
	0	1	-0.04	4.76
	0	1	0	1.00
	0	1	0.04	0.21
	0	1	0.08	0.04
	0	1	0.12	0.01
	0	1	0.16	0.00
	0	1	0.2	0.00

b.

$\Delta G^{\circ\prime}$	Charge	$\Delta\Psi_{cell-med}$	$[S]_{cell}/[S]_{med}$
0	2	-0.2	5910040.51
0	2	-0.16	261380.01
0	2	-0.12	11559.91
0	2	-0.08	511.25
0	2	-0.04	22.61
0	2	0	1.00
0	2	0.04	0.04
0	2	0.08	0.00
0	2	0.12	0.00
0	2	0.16	0.00
0	2	0.2	0.00

c.

$\Delta G^{\circ\prime}$	Charge	$\Delta\Psi_{cell-med}$	$[S]_{cell}/[S]_{med}$
0	-1	-0.2	0.00
0	-1	-0.16	0.00
0	-1	-0.12	0.01
0	-1	-0.08	0.04
0	-1	-0.04	0.21
0	-1	0	1.00
0	-1	0.04	4.76
0	-1	0.08	22.61
0	-1	0.12	107.52
0	-1	0.16	511.25
0	-1	0.2	2431.06

d.

$\Delta G^{\circ\prime}$	Charge	$\Delta\Psi_{cell-med}$	$[S]_{cell}/[S]_{med}$
0	0	-0.2	1.00
0	0	-0.16	1.00
0	0	-0.12	1.00
0	0	-0.08	1.00
0	0	-0.04	1.00
0	0	0	1.00
0	0	0.04	1.00
0	0	0.08	1.00
0	0	0.12	1.00
0	0	0.16	1.00
0	0	0.2	1.00

Experiment 7.1

1.

	$d = 1.5$ Å		$d = 1.75$ Å		$d = 2.00$ Å	
ø	r	Allowed?	r	Allowed?	r	Allowed?
180	3.77	Yes	4.39	Yes	5.02	Yes
135	3.61	Yes	4.21	Yes	4.81	Yes
90	3.19	Yes	3.72	Yes	4.25	Yes
45	2.70	No	3.15	Yes	3.61	Yes
0	2.48	No	2.89	No	3.30	Yes
-45	2.70	No	3.15	Yes	3.61	Yes
-90	3.19	Yes	3.72	Yes	4.25	Yes
-135	3.61	Yes	4.21	Yes	4.81	Yes
-180	3.77	Yes	4.39	Yes	5.02	Yes

2. The range of ø increases.

3. The range of ø increases.

4. 120°

5. Of the nine values of ø sampled by the model described in Figure 7.5, 4/9 of the range is allowed when $r_1 + r_2 = 3.4$ and 2/9 when $r_1 + r_2 = 3.7$. Examination of the actual values of r shows that ø is highly restricted at $r_1 + r_2 = 3.7$ at values near ±180°.

Thus increasing the size of side groups in a molecule can lock the molecule into a very narrow range of conformations.

Experiment 7.2

1. 74/81 of the combinations are allowed.
As d increases, the number of allowed combinations also increases.
As $r_1 + r_2$ decreases, the number of allowed combinations also increases.
As θ decreases, the number of allowed combinations also decreases.

Experiment 7.3

1.

r	Dielectric Constant		
	10	40	80
2	16.60	4.15	2.08
4	8.30	2.08	1.04
6	5.53	1.38	0.69
8	4.15	1.04	0.52
10	3.32	0.83	0.41
12	2.77	0.69	0.35
14	2.37	0.59	0.30
16	2.08	0.52	0.26
18	1.84	0.46	0.23

2.

Dist.	Electrostatic Potential	Lennard-Jones Potenial
2.0	2.08	649.1
3.0	1.38	3.27
4.0	1.04	-0.18
5.0	0.83	-0.08
6.0	0.69	-0.03

3.

Ionic Strength	10 Å	30 Å	100 Å
0	0.41	0.14	0.04
0.001	0.37	0.10	0.01
0.01	0.30	0.05	0.00
0.1	0.15	0.01	0.00

a. Increasing ionic strengh should weaken the DNA repressor interaction.

b. Increasing ionic strength weakens the flavodoxin-resin interaction.

Experiment 7.4

1.

ΔG_t

					pH						
r	2	3	4	5	6	7	8	9	10	11	12
15	84.09	72.89	27.39	0.32	-3.60	-6.84	-7.96	-3.11	8.68	13.99	27.36
25	47.26	40.53	13.23	-3.01	-5.36	-7.30	-7.98	-5.06	2.01	5.19	13.22
50	19.63	16.27	2.62	-5.50	-6.68	-7.65	-7.99	-6.53	-2.99	-1.40	2.61

Fraction in Native Form

					pH						
r	2	3	4	5	6	7	8	9	10	11	12
15	0.00	0.00	0.00	0.37	1.00	1.00	1.00	0.99	0.00	0.00	0.00
25	0.00	0.00	0.00	0.99	1.00	1.00	1.00	1.00	0.03	0.00	0.00
50	0.00	0.00	0.01	1.00	1.00	1.00	1.00	1.00	0.99	0.91	0.01

a. No change -- at minimum $q = 0$ and $\Delta G_{ew} = 0$ independent of r_p.

b. No change in ΔG_t ($q = 0$ at pH of maximum stability)

c. $r = 25$ Å, pH 4.95 - 9.29; $r = 35$ Å, pH 4.68 - 9.56

2.

# lys	\multicolumn{11}{c}{pH}
	2	3	4	5	6	7	8	9	10	11	12
4	18.30	13.73	-2.70	-8.00	-7.54	-5.91	-2.93	0.21	3.38	5.36	13.24
8	31.45	25.80	3.94	-6.83	-7.78	-7.91	-6.55	-2.76	2.68	5.28	13.23
12	47.26	40.53	13.23	-3.01	-5.36	-7.30	-7.98	-5.06	2.01	5.19	13.22

a. pH of maximum stability increases.

b. ΔG_t increases (protein becomes less stable) at low pH (charge in protein increases)

c. ΔG_t changes very little (lysines are uncharged)

3.

# asp	\multicolumn{11}{c}{pH}
	2	3	4	5	6	7	8	9	10	11	12
8	47.43	42.00	18.87	2.78	-0.35	-4.05	-7.01	-7.68	-3.95	-1.85	3.93
12	47.26	40.53	13.23	-3.01	-5.36	-7.30	-7.98	-5.06	2.01	5.19	13.22
16	47.09	39.08	8.25	-6.59	-7.77	-7.90	-6.29	0.21	10.63	14.89	25.16

a. pH of maximum stability decreases.

b. ΔG_t changes very little (aspartates are uncharged).

c. ΔG_t increases at high pH values (charge on protein increases).

Experiment 7.5

1. 3
2. 8

INDEX

2-aminobenzoic acid 60
3-PO$_4$-glyceraldehyde 220, 223
3-PO$_4$-glycerate 223
3-PO$_4$-glyceraldehyde 167
α-ketoglutarate 223
absolute 16, 17, 19, 21, 22, 37, 68, 157, 161, 162, 164, 274, 311
acetaldehyde 220, 223
acetamide 233
acetyl-CoA 223, 224
acid 3, 14, 16, 19, 21, 25, 38-41, 49, 52-56, 59-61, 64, 102, 103, 106, 197, 199, 209, 243, 246, 267, 268, 291, 292, 295, 298, 302, 305-309
acidic dissociation 38
activation 68
active transport 228, 253, 256, 258, 260
activities 166, 188
activity 36, 79, 80, 90, 103, 105-107, 194
adaptation 217
adenine 6-8
adenylate 218
ADP 210, 211, 213, 218, 219, 256-261, 265
aerobic 219, 224
alanine 60, 308
alcohols 303
aldolase 170
aldolase-catalyzed 167, 170
alkane 303
alkyl 303
amide 310
amine 60
amino acid 19, 25, 41, 60, 197, 209, 243, 246, 267, 268, 295, 298, 302, 308, 309
ammonia 61-64
ammonium 61
anaerobic 224
analog 157-160, 162-164
aniline 60
anion 264
anthranilic acid 60
aqueous 3, 55, 56, 58-60, 229, 289, 291-293, 303, 304, 306, 307
arginine 41-43, 45-48, 174
arginine ionic forms 43, 46, 47
aromatic 303
Arrhenius 68
aspartate 24, 158, 300
aspartate transcarbamylase 158
ATP 36, 209-211, 213-215, 217-219, 223, 256-260, 265
ATP-linked 223

ATPase 265
bacteria 224, 258
bacterial 237
barrier 61-64
base 6, 7, 11-14, 16, 21, 39-41, 49, 52, 53, 55, 56, 59-61, 64, 102, 103, 291, 292
base pair 7
benzoic 60
bilayer 227, 229, 260, 306, 307
binding 97, 98, 100, 109, 110, 112, 113, 115, 123, 125, 129, 131-133, 135-137, 139, 141, 148, 149, 153-155, 194, 196, 208, 217, 288
blood 61-64, 233
blood-urine 62
Boltzmann 68, 70, 274
bond 139, 267-269, 272-278, 281, 310
Briggs 67
Briggs-Haldane 73-78
buffer 49, 51-54
butyramide 233
carbohydrate 224
carboxyl 306, 307
carboxylic 55, 60, 292
carrier-mediated 248, 251
catabolic 190
catalysis 131
catalyst 183
cation 262, 263
cellular 36, 217, 247, 265
chloroplasts 228
citrate 158, 224
cleavage 167, 170
cleaved 168
columnwise 26-28, 33, 151, 174, 200
competition 91, 92, 155, 190
competitive 92, 93, 155, 161, 163, 164
competitive inhibition 161
complementary 6, 8
complex 1, 19, 31, 61, 72, 73, 79, 91, 93, 122-125, 153, 188, 199, 218, 238
configurational entropy 311
conformation 296, 301, 308, 311
conjugate acid 16, 21, 39, 41, 49, 52, 55, 56, 59-61, 64, 102, 103, 291, 292
conjugate base 16, 21, 39, 41, 49, 52, 55, 56, 59-61, 64, 102, 103, 291, 292
convergence 207
cooperative 67, 86, 109, 110, 112, 113, 115, 117, 123, 124, 130-132, 136, 137, 139-142, 146, 148, 149, 152-157, 162-164, 216
cooperative kinetics 109, 131, 141, 149, 157, 216

343

Index

cooperativity 109, 112, 113, 115, 116, 131, 136-139, 147, 153, 158, 164, 191, 198, 208
covalent 72, 77
cysteine 24, 48
cytosine 6, 8
Debye 285
dehydrogenase 158, 223
dielectric 283-286, 290, 293, 294
diffusion 228, 234, 239, 241
dihedral 267, 269, 282
dihedral angle 269
dihydroxyacetone-PO4 167
dipolar 60
dissociable 62
dissociation 38, 41, 66, 67, 72-74, 76, 78, 91, 109, 117, 121, 124-126, 131-133, 149, 157, 159, 211, 212, 242
distribution 55, 58, 60, 61, 68, 130, 139, 153, 256, 258, 260, 265, 274, 289-292
distribution of ammonia 61
divalent 263
DNA 6-8
donor 105, 107, 310
effective K_s 117, 118, 121, 132
electrical field 260
electrochemical 219, 226, 260
electrode 220-223, 225, 226
electron 219, 221, 225
electrostatic 266, 283, 284, 286-292, 294, 295, 301
endoplasmic reticulum 228
energy-transducing 209
enthalpy 311
entropic 310
entropy 310, 311
enzyme 54, 65, 66, 69, 72, 78, 79, 81, 82, 86, 88-93, 96, 101, 102, 105-107, 109, 110, 112, 113, 117, 118, 121, 123-126, 129-133, 135-139, 146, 148, 149, 152-158, 162, 170, 184, 186-190, 193, 194, 197, 198, 204, 207, 208, 212, 214-217, 238-240
enzyme kinetics 65, 69
enzyme-catalyzed 65, 72, 76, 91, 92, 106, 141, 147, 179
enzyme-ligand 72
enzyme-product 91, 93
enzyme-substrate 73, 78, 91, 93, 153
equilibria 48, 91, 172, 174
equilibrium 31, 32, 34, 36, 38, 64, 66, 67, 72-74, 91, 92, 100, 109, 110, 124, 126, 133, 165-170, 172-174, 176-178, 181, 209-211, 219, 224, 225, 237, 253, 257, 258, 260, 261, 304
ethanol 220, 223, 251, 307
ether 60

E_{tot} 150, 151
eucaryotic 227, 228
exponential 68, 285
extracellular 238, 246, 265
Eyring 68
Eyring kinetics 68
facilitated diffusion 228, 241
facilitated transport 248, 249
Faraday 219, 260
fat 209
feedback 100, 140, 197-199, 201, 205-208
feedback control 197, 199, 207
flavoprotein 219, 221, 224
flow 31, 178, 194, 202, 219, 227, 239
flux 1, 177, 178, 180-184, 186-194, 198, 201, 202, 204-207, 212, 216, 217, 224, 228, 230, 239-242
force 28, 35, 200, 283, 311
forces 1, 13, 49, 266, 287, 290, 292, 296, 310-312
forward reference 26, 27, 201
fructokinase 210
fructose 167, 168, 170
fructose-6-PO$_4$ 210, 211
fumarate 224
fusion 228
futile-cycling 215
Gibbs 172, 217, 219, 221
gluconeogenesis 209, 223
glucose 36, 171
glucose-6-PO$_4$ 36
glutamine 61
glycerol 233-235, 238, 250
glycerol-1-PO$_4$ 258
glycine 264, 268
glycolysis 171, 209, 223
gradient 64, 260
GTP 223
guanine 6-8
Haldane 67
half-time 237, 244, 245
helical 6
heme 148
hemoglobin 148
Henderson-Hasselbalch equation 3, 14, 17, 19, 23, 38, 44, 49, 56, 61, 62, 103, 292
heterotrophic 219
high-order 116
Hill 141-148
Hill equation 141, 143, 145
Hill plot 141, 142
histidine 48
homeostasis 231
homocysteine 246

Index

human 232-234, 238, 243, 244, 250, 251
hydrogen 6, 139, 267, 310, 311
hydrolysis 61, 215, 256-258, 260
hydrolyzed 209
hydrophilic 227, 303
hydrophobic 227, 266, 303, 306-308, 310
hydrophobic effect 303, 308, 310
hygiene 34
impermeable 62, 227
inhibition 92, 96-99, 157, 161, 190, 191
inhibitor 92, 93, 95, 97, 98, 101, 161, 162
inorganic 36, 258
intercalate 227
interconversion 91, 224
intermediary 197, 199
intermediate 3, 184, 187, 188, 223
intracellular 230, 233-235, 237, 253
intrinsic 125, 131, 143, 149, 208
ionic 42-48, 60, 106, 260, 285, 287, 288, 301
ionic solute 260
ionizable 14, 19, 42, 106, 289-291, 295
ionization 19, 42, 48, 174, 293
ionizing 25, 43, 48
isocitrate 158, 223
isoelectric 46-48
isoelectric point 47, 48
isomerase 170, 171
iterate 204
iterating 202
iteration 199-201, 207
K_{cat} 139
K_{eq} 31-33, 35, 37, 94, 96, 100, 101, 166-170, 183, 212, 213, 215, 217, 219, 221-224, 226
kidney 61
kinase 223
kinetic 67, 72, 80-82, 84, 86, 92, 96, 107, 109, 113, 115, 129, 141, 143, 145, 146, 157, 177, 184, 206, 214, 215, 217, 250
K_{int} 125-128, 130, 131, 133, 134, 139, 143, 144, 148-151, 153-155
K_m 66, 67, 73, 74, 78, 83, 84, 86-90, 129, 130, 132, 156, 184-189, 199, 204, 205, 207, 217, 224, 239, 240, 242, 244-247, 251, 252
K_s 66, 67, 73, 97, 109, 110, 117-121, 132, 158, 216
K_s' 109, 117, 119-121
lactate 223, 224
leucine 308
liberate 226
ligand 76, 92, 109, 124, 126, 129, 131-133, 135-137, 148, 157, 158, 161, 162, 242
ligand-bound 157
ligand-receptor 125
linear 67, 80, 82, 86, 176

Lineweaver-Burk 155
lipid 227, 307, 310
lipid bilayer 227, 307
liver 224
loop 198, 201, 208
lysine 47, 299, 300
lysosome 228
macromolecular 197, 283, 294, 303, 310, 311, 313
macromolecule 266, 303, 310-312
malate 223, 224
mammalian 236
mechanism 132, 241, 251, 252
mediated 223
membrane 1, 61, 64, 227-230, 232, 234, 236, 238, 239, 241, 247, 250, 253, 258, 260, 262, 306, 307
metabolic 1, 31, 109, 113, 115, 130, 171, 176, 179, 184, 186, 188, 196, 197, 206, 209, 215, 217-219, 223, 224, 236, 237, 240
metabolic pathway 1, 184, 188, 197
metabolism 31, 149, 165, 171, 172, 178, 190, 209, 217-219, 221, 224, 226
metabolite 188, 190, 194, 196, 197, 199, 201, 205, 223, 240
methionine 243-247, 308
methyl-donor 246
methylation 246
Michaelis 65-67, 72-82, 86, 88, 91, 93, 117, 118, 127-132, 143, 152, 157, 185, 198, 201, 238, 242
Michaelis-like 76
Michaelis-type 211
microorganisms 224
midpoint 49, 54, 147, 148
migrating 262-264
mitochondria 228
mitochondrial 158
modifier 113, 117, 120-122, 149, 152-156, 194, 196, 198, 208
moment 157
momentary 80-82, 84, 86, 143, 144, 197, 209
Monod-Wyman-Changeaux 148
monovalent 262, 264
multiple sequential equilibria 48, 172, 174
multisite 112
mutation 100
NAD 219-224
NADH 220, 224
negative cooperativity 153
neurotransmitter 228
nitrogen 310
non-cooperative 141
non-interacting 129

Index

non-interacting sites 129
noncooperative 138, 141-143, 146, 153, 162
noncovalent 266
nonpolar 55, 231, 289-291, 293, 303, 304, 306-308, 310
nucleic acid 199
nucleus 228
O_2 219, 221-223, 227
one-product 65, 72, 91
one-substrate 65, 72, 91
optimize 193
order 15, 26-29, 33, 80-84, 86, 109, 115, 140-148, 151, 157, 158, 174, 175, 179, 200, 206, 211, 213, 214
organelles 228
oxaloacetate 224
oxidation 209, 219, 221, 223
oxidation-reduction 225
oxidative 224
oxidized 225, 226
oxidizing 219, 220, 223, 224, 226
oxygen 141, 224, 310
oxygen-hemoglobin 148
partition 44, 45, 55, 56, 92, 124, 130, 173, 291, 304
partition between phases 55
partition equation 56, 92, 124, 130
passive 228, 229, 239, 248, 249, 251
pathway 1, 31, 42, 184, 188, 190, 191, 193, 194, 196, 197, 199, 201, 206
permeability 227, 229, 238
permeable 61
peroxisome 228
pH 2-4, 9, 14, 16-18, 20-24, 39-41, 43-49, 51-55, 59-64, 102-108, 220, 226, 289, 292-295, 298-302
pH dependence 294
pH response 102, 106, 107
phase 55, 56, 58-60, 291, 293, 303, 304
phenylalanine 308
phosphatase 210, 215
phosphate 36, 170, 210, 265
phosphoenolpyruvate 258
phosphofructokinase 211, 215
phosphoglyceraldehyde 171
phospholipid 227, 306
phospholipid-cholesterol 227
phosphorylating 36
physiological 46, 60
pK 3, 4, 14, 16-18, 21, 22, 24, 38-41, 49, 52, 54, 58, 60, 62, 103-108, 292, 293, 295, 300
plasma 227, 246, 247
PO_4 258, 259
polar 55, 227, 288, 290, 291, 293, 303, 304, 306, 311
polarity 55
polynucleotide 6, 8
polypeptide 9, 25, 267, 281
potassium 265
potential 2, 27, 219-222, 226, 253, 256, 260, 261, 283-288
precursor 178, 180
principal 45, 242, 294
probability 102, 131, 137
probe 3
procaryotic 227
protein 19, 197, 199, 227, 241, 246, 247, 266, 268, 275, 294-296, 298, 299, 301, 302, 307, 308, 313
protein-CH_3 246
proton 54, 102, 105, 107
proton-accepting 102, 103, 106
proton-donating 102, 103, 106
protonated 47, 48
prototype 18, 21, 22, 57, 94, 127, 134, 144, 160, 193, 297
push-pull 214
pyruvate 223, 224
quadratic 32, 191
quadratic equation 32, 191
reaction rate 70, 76, 80, 86, 90, 91, 96, 102, 107, 110, 112, 115, 153, 157, 158, 161-164, 181, 184, 200
reagent 34, 35
receptor 109
redox 219
redox reactions 219
reducing 226
regeneration 223
regulation 86, 149, 198, 217
regulator 113
regulatory 109, 113-117, 122, 123, 154, 184, 188, 193, 194, 197-199, 208, 209, 216, 217
renal 61, 62, 64
replicate 16, 17
residues 23, 24, 42, 267, 299, 300, 302, 308, 311
ribonuclease 308
ribose-5-PO_4 36
rowwise 26, 28
s-adenosylmethionine 246
saturate 92
second-order 80, 211, 216, 217
secrete 61, 62
sequential 48, 172, 174, 177
sequential reactions 177
slope 54, 80-84, 141-144, 148, 155, 206
sodium 265, 286-288
sodium-potassium 265

Index

solubilities 56, 291, 292, 303
solubility 59, 60, 292, 303
soluble 55, 59, 227, 303, 305
solute 59, 229, 230, 237, 241, 242, 253, 255-258, 260
solvent 172, 303
stabilization 206, 217, 307, 308
stabilize 311
steady-state 36, 188, 199, 202, 239, 246, 247
stoichiometries 209
stoichiometry 165, 217, 258, 265
storage 30, 209
strand 6-8
subcellular 228
substituted 182, 237, 243
substrate affinity 109, 117
substrate-binding 113
substrate-enzyme 148
succinate 158, 223, 224
sugar 170
surface 80, 230, 233-235
symmetrical 147, 148, 215, 242
symmetry 64, 275
syntheses 197
synthesis 35, 197, 207, 209
synthetase 224
synthetic 190
terminal 19

thermodynamic 67, 209, 217, 224
thiokinase 223
three-dimensional structure 267, 268
thymine 6, 8
titration 47
torsion angle 268, 269, 275, 278
transcarbamylase 158
transport 1, 227-231, 236, 238, 242-244, 246, 248-253, 256-258, 260, 265
triose 170
triose-PO$_4$-isomerase 171
tryptophan 308
tubule 61, 62, 64
tyrosine 24, 48, 308
urine 61-64
valine 308
van der Waals 274, 276
velocity 65-68, 71, 88, 105, 108, 110, 114, 115, 117, 118, 121, 132, 140, 141, 184, 193
V_{max} 66, 67, 79, 83, 84, 86-89, 91, 92, 111-113, 115, 117, 119, 121, 132, 141, 143, 144, 150, 151, 153-155, 184-189, 191-196, 199, 204, 205, 207, 208, 211-213, 215, 217, 239, 240
V_{net} 215, 216
volt 219, 220, 260
yeast 158
zwitterion 264